Springer Theses

Recognizing Outstanding Ph.D. Research

Aims and Scope

The series "Springer Theses" brings together a selection of the very best Ph.D. theses from around the world and across the physical sciences. Nominated and endorsed by two recognized specialists, each published volume has been selected for its scientific excellence and the high impact of its contents for the pertinent field of research. For greater accessibility to non-specialists, the published versions include an extended introduction, as well as a foreword by the student's supervisor explaining the special relevance of the work for the field. As a whole, the series will provide a valuable resource both for newcomers to the research fields described, and for other scientists seeking detailed background information on special questions. Finally, it provides an accredited documentation of the valuable contributions made by today's younger generation of scientists.

Theses are accepted into the series by invited nomination only and must fulfill all of the following criteria

- They must be written in good English.
- The topic should fall within the confines of Chemistry, Physics, Earth Sciences, Engineering and related interdisciplinary fields such as Materials, Nanoscience, Chemical Engineering, Complex Systems and Biophysics.
- The work reported in the thesis must represent a significant scientific advance.
- If the thesis includes previously published material, permission to reproduce this must be gained from the respective copyright holder.
- They must have been examined and passed during the 12 months prior to nomination.
- Each thesis should include a foreword by the supervisor outlining the significance of its content.
- The theses should have a clearly defined structure including an introduction accessible to scientists not expert in that particular field.

More information about this series at http://www.springer.com/series/8790

Yuichi Hirai

Assembled Lanthanide Complexes with Advanced Photophysical Properties

Doctoral Thesis accepted by
the Hokkaido University, Hokkaido, Japan

Author
Dr. Yuichi Hirai
Hokkaido University
Sapporo, Hokkaido
Japan

Supervisor
Prof. Yasuchika Hasegawa
Graduate School of Engineering
Hokkaido University
Sapporo, Hokkaido
Japan

ISSN 2190-5053   ISSN 2190-5061   (electronic)
Springer Theses
ISBN 978-981-10-8931-2   ISBN 978-981-10-8932-9   (eBook)
https://doi.org/10.1007/978-981-10-8932-9

Library of Congress Control Number: 2018936640

© Springer Nature Singapore Pte Ltd. 2018
This work is subject to copyright. All rights are reserved by the Publisher, whether the whole or part of the material is concerned, specifically the rights of translation, reprinting, reuse of illustrations, recitation, broadcasting, reproduction on microfilms or in any other physical way, and transmission or information storage and retrieval, electronic adaptation, computer software, or by similar or dissimilar methodology now known or hereafter developed.
The use of general descriptive names, registered names, trademarks, service marks, etc. in this publication does not imply, even in the absence of a specific statement, that such names are exempt from the relevant protective laws and regulations and therefore free for general use.
The publisher, the authors and the editors are safe to assume that the advice and information in this book are believed to be true and accurate at the date of publication. Neither the publisher nor the authors or the editors give a warranty, express or implied, with respect to the material contained herein or for any errors or omissions that may have been made. The publisher remains neutral with regard to jurisdictional claims in published maps and institutional affiliations.

Printed on acid-free paper

This Springer imprint is published by the registered company Springer Nature Singapore Pte Ltd. part of Springer Nature
The registered company address is: 152 Beach Road, #21-01/04 Gateway East, Singapore 189721, Singapore

Supervisor's Foreword

Luminescent materials are vital for human activities, and the development of novel luminescent compounds is important to enrich our lives as well as to explore scientific principles. In particular, there has been significant recent interest in luminescent lanthanide(III) complexes for applications such as optical materials, organic light-emitting diodes (OLEDs), and fluorescent sensors due to their line-like, long-lived, and color-tunable emission.

It has been 180 years since the first emission of an europium(III) complex was described; however, there still remains a plenty of room for improvement and much potential for functional luminescent materials because of the large coordination numbers and flexible coordination geometry of lanthanide(III) ions. In addition, these "rare earth" metals are actually not that rare, compared to other familiar transition metals such as ruthenium, gold, and platinum. These features make lanthanides more attractive as promising candidates for future luminescent materials.

Due to the low excitation coefficients of lanthanide(III) ions, organic chromophores that can serve as strong absorbing "antenna ligands" have been developed. Typical antenna ligands including beta-diketonates and pyridine derivatives have tens of thousands of larger molar absorption coefficients than lanthanide(III) ions, and the absorbed energy populates the 4f excited states through ligand-to-metal energy transfer, resulting in intense lanthanide(III)-centered emission. The photophysical properties of these lanthanide(III) complexes are dominated by symmetry and vibration of coordination structures. In the previous works, asymmetrical and low-vibrational coordination structures were constructed to increase the probability of parity-forbidden 4f-4f transitions and to suppress vibrational relaxation processes.

In the recent two decades, lanthanide(III)-based coordination polymers (CPs) and metal organic frameworks (MOFs) have been significantly studied for additional physical properties. Although lanthanide(III)-based MOFs provide thermal stability, color tunability, and ion/molecular sensitivity, it is generally difficult to incorporate ideal antenna ligands into the rigid frameworks. On the other hand, one-dimensional CPs let antenna ligands to coordinate, and have more structural flexibility due to the lack of three-dimensional networks. Therefore, lanthanide(III)-based CPs are expected as highly luminescent materials with additional advanced properties by controlling the assembled structures.

In this book, assembled lanthanide(III) complexes with advanced photo, thermal, and mechanical properties are demonstrated by Dr. Yuichi Hirai. YH here adopted one of the most common antenna ligands (hexafluoroacetylacetonato: hfa) and simply focused on bridging ligands to control entire structures. YH has considered three points, namely (i) densely packed structures for high energy transfer efficiency (η_{sens}), (ii) loose-packed structures for glassy-state luminescence, and (iii) mechanically unstable structures for fracture-induced luminescence.

YH made a start with developing lanthanide(III) CPs with high energy transfer efficiency. There are many reports on extremely high intrinsic quantum yields (> 90%), but it is still challenging to achieve high η_{sens} in assembled systems. In 2008, a group of Switzerland reported that intra-ligand charge transfer (ILCT) state enhanced the energy transfer of a Eu(III) dimer complex with close-packed hfa ligands in solid state. Inspired by this idea, YH synthesized thiophene-based bridging ligands since small π-aromatic cores and bent coordination angles were expected to tighten each unit of CPs. The resulting compounds had densely packed structures with twice as large η_{sens} as previous reports, and multiple intermolecular hydrogen bonds also provided thermal stability. The energy transfer efficiencies between Tb(III) and Eu(III) ions were also studied using lanthanide(III)-mixed CPs for temperature-sensitive materials. The emission color and temperature sensitivity were found out to be tunable by Tb(III)/Eu(III) ionic ratios. These efficient and color-tunable luminophores are highly expected as industrious applications.

On the contrary to the densely packed CPs, loose-packed amorphous compounds are ideal for luminescent paintings on materials' surface. A series of C_3 symmetrical organic compounds with T_g at ambient temperatures, so-called amorphous molecular materials, has various uses due to the advantages arising from "amorphous" and "molecular" properties such as transparency, homogeneity, and purity without M_w distributions. YH came up with the idea that C_3 symmetrical structures would be formed by connecting 8-coordinated monomer lanthanide(III) units with 120-degree-bent bridging ligands. Powder XRD patterns clearly demonstrated the noncrystalline states, and multiple *quasi*-stable structures in DFT optimization process indicated thermodynamically nonequilibrium states. The ethynyl groups in the bridging ligands are also found to play an important role in suppressing tight-packing by enlarging the cell volume without extending π-aromatic surface for intermolecular interactions. Tb(III)/Eu(III) mixed complexes also exhibited temperature sensitivity in glassy state.

As a final topic, YH proposed a systematic molecular design for fracture-induced luminescence (triboluminescence: TL), which is based on relatively loose-packed but crystalline structures. TL has been widely found in crystalline materials since the glow of sugar was noted in the seventeenth century, and also known as a storm in a teacup due to its puzzling emission mechanisms. Still now, it is controversial if the TL needs non-centrosymmetrical space groups, what can be an excitation source, and what is the difference between well-studied photoluminescence (PL).

YH found that lanthanide(III) compounds with face-to-face arrangements exhibited strong TL. Surprisingly, he observed clearly different emission spectra of TL and PL in Tb(III)/Eu(III) mixed systems, being the first example of direct evidence of their distinct excitation and/or energy transfer processes. Although quantitative measurements are required for further understanding, these results would break new ground of mechano-luminescent systems for both academic and industry.

In this book, luminescent lanthanide(III) complexes with specific assembled structures are introduced by Dr. Yuichi Hirai. The proposed molecular designs are significantly expandable for novel classes of solid-state luminescent materials. I'm happy that a series of these works is being published as a book to attract readers into the science of lanthanide(III) luminescence.

Hokkaido, Japan Prof. Yasuchika Hasegawa
January 2018

Parts of this thesis have been published in the following journal articles:

- Y. Hirai, T. Nakanishi, K. Miyata, K. Fushimi, Y. Hasegawa, "Thermo-sensitive luminescent materials composed of Tb(III) and Eu(III) complexes," *Mater. Lett.*, **2014**, *130*, 91–93.
- Y. Hirai, T. Nakanishi, Y. Kitagawa, K. Fushimi, T. Seki, H. Ito, H. Fueno, K. Tanaka, T. Satoh, Y. Hasegawa, "Luminescent Coordination Glass: Remarkable Morphological Strategy for Assembled Eu(III) Complexes," *Inorg. Chem.*, **2015**, *54*, 4364–4370.
- Y. Hirai, T. Nakanishi, Y. Kitagawa, K. Fushimi, T. Seki, H. Ito, Y. Hasegawa, "High energy transfer efficiency of Eu(III) coordination polymers with thiophene-derivative bridge," *Angew. Chem. Int. Ed.*, **2016**, *55*, 12059–12062.
- Y. Hirai, P. P. Ferreira da Rosa, T. Nakanishi, Y. Kitagawa, K. Fushimi, Y. Hasegawa, "Amorphous Formability and Temperature-Sensitive Luminescence of Lanthanide Coordination Glasses Linked by Thienyl, Naphthyl, and Phenyl Bridges with Ethynyl Groups," *Bull. Chem. Soc. Jpn.*, **2017**, *90*, 322–326.
- Y. Hirai, T. Nakanishi, Y. Kitagawa, K. Fushimi, T. Seki, H. Ito, Y. Hasegawa, "Triboluminescence of Lanthanide Coordination Polymers with Face-to-Face Arranged Substituents," *Angew. Chem. Int. Ed.*, **2017**, *56*, 7171–7175.

Acknowledgements

First and foremost, I would like to express my sincere gratitude to my advisor Yasuchika Hasegawa for his continuous support of my Ph.D. study and related research over a period of 5 years. Besides my advisor, I would like to thank Hiroki Habazaki, Masako Kato, Koji Fushimi, Hajime Ito, and Toshifumi Satoh to be my examiners and to give a lot of advice and encouragement. I also thank Takayuki Nakanishi, and Yuichi Kitagawa for their kind help, suggestions, and fruitful discussions.

It is a pleasure to thank Kazuyoshi Tanaka, Hiroyuki Fueno, Keiji Morokuma, and Miho Hatanaka for collaborative work on quantum calculations of coordination structures and energy transfer efficiencies using a cutting-edge supercomputer and theories.

I express my thanks to RIGAKU Co. and SHIMADZU Co. for single-crystal X-ray analyses of micro-sized crystals, DSC measurements, and SPM observations.

I would like to acknowledge the support by the Japan Society for the Promotion of Science (Grants-in-Aid for JSPS fellows), Grants-in-Aid for Scientific Research on Innovative Areas of "New Polymeric Materials Based on Element-Blocks," and Frontier Chemistry Center Akira Suzuki "Laboratories for Future Creation" Project.

Finally, and most importantly, I would like to thank my parents Mitsuhiro Hirai and Rika Hirai for their understanding and support.

Contents

1 **General Introduction** 1
   1.1 Luminescent Materials 1
   1.2 Luminescence of Lanthanide(III) Ions 3
   1.3 Enhancement of Luminescence Efficiency 3
      1.3.1 The Antenna Effect 3
      1.3.2 Promotion of 4f-4f Transitions 5
      1.3.3 Suppression of Vibrational Relaxation 7
   1.4 Assembled Structures of Coordination Compounds 8
   1.5 Assembled Ln(III) Coordination Compounds 9
   1.6 Objectives 10
   1.7 Contents of This Thesis 10
   References 11

2 **Luminescent Lanthanide Coordination Zippers with Dense-Packed Structures for High Energy Transfer Efficiencies** 15
   2.1 Introduction 15
   2.2 Experimental Section 17
      2.2.1 General 17
      2.2.2 Apparatus 17
      2.2.3 Syntheses 18
      2.2.4 Crystallography 20
      2.2.5 Assignment of Coordination Geometry 20
      2.2.6 Optical Measurements 21
      2.2.7 Estimation of Intrinsic Emission Quantum Yields .. 21
   2.3 Results and Discussion 22
      2.3.1 Photophysical Properties 22
      2.3.2 Coordination Structures 24

		2.3.3	Thermal Properties	27
		2.3.4	DFT Calculations	30
	2.4	Conclusions		30
	References			32
3	**Luminescent Lanthanide-Mixed Coordination Polymers for Tunable Temperature-Sensitivity**			**35**
	3.1	Introduction		35
	3.2	Experimental Section		38
		3.2.1	Materials	38
		3.2.2	Apparatus	38
		3.2.3	Syntheses	38
		3.2.4	Optical Measurements	41
	3.3	Results and Discussion		41
		3.3.1	Powder X-Ray Diffraction Measurements	41
		3.3.2	Emission Spectra	41
		3.3.3	Energy Transfer Efficiency	43
	3.4	Conclusions		44
	References			45
4	**Luminescent Lanthanide Coordination Glasses**			**47**
	4.1	Introduction		47
	4.2	Experimental Section		49
		4.2.1	Materials	49
		4.2.2	Apparatus	49
		4.2.3	Syntheses	49
		4.2.4	Crystallography	52
		4.2.5	Optical Measurements	53
		4.2.6	DFT Calculations	53
	4.3	Results and Discussion		53
		4.3.1	Structural Characterization	53
		4.3.2	DFT Optimization	56
		4.3.3	Thermal Properties	58
		4.3.4	Surface Observations by Scanning Probe Microscope (SPM)	61
		4.3.5	Photophysical Properties	63
	4.4	Conclusions		64
	References			66
5	**Amorphous Formability and Temperature-Sensitive Luminescence of Lanthanide Coordination Glasses**			**69**
	5.1	Introduction		69
	5.2	Experimental Section		71
		5.2.1	General	71

		5.2.2	Apparatus	71
		5.2.3	Syntheses	71
		5.2.4	Optical Measurements	74
	5.3	Results and Discussion		74
		5.3.1	Structural Characterizations	74
		5.3.2	Thermal Properties	75
		5.3.3	Photophysical Properties	77
	5.4	Conclusions		78
	References			79
6	**Triboluminescence of Lanthanide Coordination Polymers**			81
	6.1	Introduction		81
	6.2	Experimental Section		82
		6.2.1	General	82
		6.2.2	Apparatus	83
		6.2.3	Syntheses	83
		6.2.4	Crystallography	85
		6.2.5	Optical Measurements	86
	6.3	Results and Discussion		86
		6.3.1	Coordination Structures	86
		6.3.2	TL Activity of Eu(III) Coordination Polymers	89
		6.3.3	Photoluminescence (PL) Properties	89
		6.3.4	PL and TL Properties in Tb(III)/Eu(III) Mixed Systems	92
		6.3.5	PL and TL Properties of Gd(III) Coordination Polymers	94
		6.3.6	Atmospheric Dependence of TL and PL	94
		6.3.7	Summary of Observed Properties and Further Assumptions	95
	6.4	Conclusions		97
	References			99
7	**Summary and Outlook**			101
	7.1	Summary		101
	7.2	Outlook		102
Curriculum Vitae				105

Chapter 1
General Introduction

Abstract Luminescent materials play a vital role in our lives. There has been significant progress in the development of inorganic and organic phosphors. Assembled lanthanide [Ln(III)] coordination compounds have attracted much attention as promising candidates for luminescent materials due to their unique photophysical properties arising from f-f transitions. In this thesis, the relationships between assembled Ln(III) coordination compounds and photophysical, thermal, and mechanical properties are focused on for the development of novel luminescent materials. The molecular designs of organic bridging ligands to control the assembled structures and physical properties of Ln(III) coordination compounds are shown.

Keywords Luminescence · Lanthanide · Complex · Coordination polymer

1.1 Luminescent Materials

The development of efficient and bright luminescent materials has broadened the range and time of activities of human beings who rely to a large extent on visual perception. After the emergence of incandescent lamps, artificial light sources have contributed greatly to our lifestyle changes [1–3]. The first commercial light-emitting diode (LED) was developed by Holonyak and Bevacqua based on GaAsP in 1962 [4]. Akasaki, Amano, and Nakamura developed the high-power blue LED using GaN [5–8]. Kido and Adachi also laid the foundation for organic- and lanthanide-based electroluminescent devices [9–11]. Numerous types of inorganic and organic luminescent materials are now used in various applications such as displays, lightings, lasers, and optical communication systems (Fig. 1.1) [12–15].

Most inorganic phosphors are composed of host crystals with wide band gaps (e.g., oxides, nitrides, and sulfides) and a small amount of dopants including rare earth and transition metal ions. The dopant ions are referred to as activators and serve as luminescent centers. The photophysical properties of these materials are tuned by varying the host/dopant combinations or the concentrations of dopants, and these materials have advantages of stability due to the rigid inorganic frameworks.

© Springer Nature Singapore Pte Ltd. 2018
Y. Hirai, *Assembled Lanthanide Complexes with Advanced Photophysical Properties*,
Springer Theses, https://doi.org/10.1007/978-981-10-8932-9_1

Fig. 1.1 Luminescent materials in daily lives

Schnick and co-workers recently developed a narrow-band and high-efficient red nitride phosphor, $Sr[LiAl_3N_4]$:Eu(II) [16]. The colloidal nanocrystalline semiconductors, so called quantum dots, are also known as attractive optoelectronic materials due to their quantum confinement effects that lead to high emission quantum yields, size-dependent emission profiles, and narrow spectral bands [17].

On the other hand, organic luminescent materials are generally aromatic compounds with extended π-electron systems as represented by perylenes and coumarins [18–20]. Physical properties of these compounds can be controlled by expansion/reduction of π-conjugation systems and modification of substituents. Taking advantage of their inexpensive raw materials with little restriction on resources, high-dispersible and easy-processable properties, they have been used in soft and flexible optical devices such as organic displays and lightings [21–23].

In addition, coordination compounds are also attractive because of their various combinations of metal ions and organic ligands. One of the most remarkable points would be their diversity of coordination structures. They have been reported to form polyhedral clusters, polymer chains, and frameworks depending on metal ions and organic ligands, which results in varied chemical and physical properties [24]. These assembled inorganic/organic systems can provide all manner of options such as sharp/broad luminescence, rigid/soft structures, and intermediate or unexpected properties that each component does not exhibit.

1.2 Luminescence of Lanthanide(III) Ions

Lanthanide(III) (Ln(III)) coordination systems composed of Ln(III) ions and organic ligands have attracted attention due to their versatile photophysical properties arising from characteristic 4f-4f transitions [25–31]. They generally show line-like and long-lived emission profiles (Fig. 1.2a) that are different from those of d-block transition metal complexes and organic compounds including metal-to-ligand charge transfer (MLCT), ligand-to-metal charge transfer (LMCT), and π-π* transitions. It is also known that the emission wavelength varies depending on Ln(III) ions from visible (VIS) to near-infrared (NIR) region (Fig. 1.2b, Table 1.1) [32, 33]. VIS emission of Tb(III), Eu(III), Dy(III), Sm(III), and Tm(III) ions has long been used for applications such as light-converting optical materials and light-emitting layers in electroluminescent (EL) devices [10, 11, 34, 35]. NIR emitters such as Yb(III), Nd(III), and Er(III) compounds have also been used in optical signal amplifiers of telecommunication networks, laser systems, and biomedical imaging sensors [36–38].

The electronic configuration of the most common +III oxidation state of Ln(III) ions is $[Xe]4f^{n-1}$, and the 4f orbitals are filled from La(III) ($4f^0$) to Lu(III) ($4f^{14}$). Since the radial charge density distribution of Ln(III) ions is smaller than the filled 5s and 5p orbitals, the 4f orbitals are shielded from the coordination environments by these orbitals. Although lanthanides bind mostly via ionic interactions, these characteristic electronic structures of Ln(III) ions provide minimal perturbation of the ligand field upon 4f orbitals. The ligand field, however, can affect the emission spectra of Ln(III) ions.

1.3 Enhancement of Luminescence Efficiency

1.3.1 The Antenna Effect

Since the 4f-4f transitions are parity-forbidden, the molar absorption coefficients of Ln(III) ions are low ($\varepsilon < 10$ M^{-1} cm^{-1}). Therefore, sensitized emission through the ligand-to-metal energy transfer (ET) is required for strong luminescence.

The sensitized emission of Ln(III) compounds was first reported by Weissman in 1942 [39]. When Eu(III) complexes with salicylaldehydes or benzoylacetonates were excited in the region of light absorption associated primarily with the ligands, strong line-like "atomic" emission from Eu(III) ions was observed. According to the findings, organic ligands with large molar absorption coefficients and a suitable energy level for ET were called "antenna ligands," and the corresponding sensitization process named "antenna effect" was confirmed (Fig. 1.3a). Then Crosby and co-workers extensively studied these mechanisms to establish the path of the energy migration process in Ln(III) complexes [40–42]. The process is explained in three steps: (i) singlet-singlet absorption of antenna ligands, (ii) intersystem crossing (ISC) from the singlet excited state (^{1}S) to the triplet excited state (^{3}T), and (iii) energy

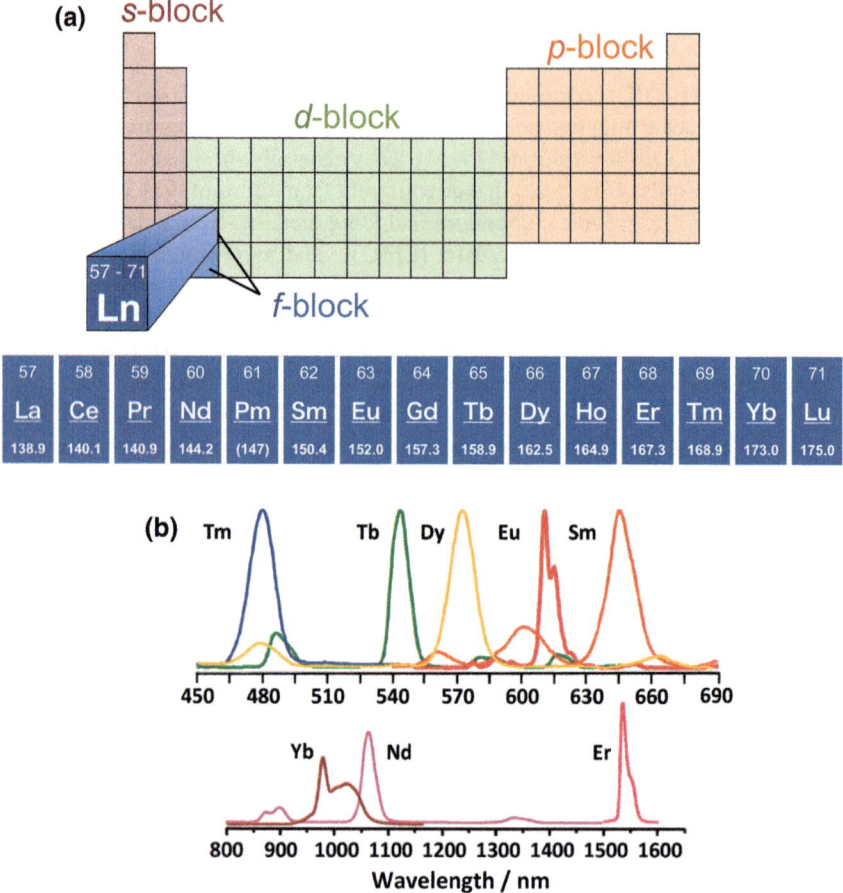

Fig. 1.2 a Lanthanide elements in periodic table, and **b** emission spectra of Ln(III) tris(β-diketonate) compounds [32]

transfer (ET) to the Ln(III) excited state, and then (iv) Ln(III)-centered emission (Em) is observed (Fig. 1.3b).

Organic chromophores such as β-diketonate and pyridine derivatives have been used as ideal antenna ligands to sensitize Ln(III)-centered luminescence [43–45]. Typical examples of β-diketonates are hexafluoroacetylacetonates (hfa) and 2-thenoyltrifluoroacetylacetonates (tta), and strong luminescence of these Ln(III)-(β-diketonates) compounds has been reported by Bünzli and others [46–48]. The group of Bettencourt also reported luminescence of Eu(III) and Tb(III) compounds with a series of pyridine-bis(oxazoline) (Pybox) derivatives [49–51].

1.3 Enhancement of Luminescence Efficiency

Table 1.1 Most common emissive f-f transitions of Ln(III) ions [33]

Element	Transition	Wavelength/nm
Pr	$^1D_2 \to {}^3F_4$	1000
	$^1D_2 \to {}^1G_4$	1440
	$^1D_2 \to {}^3H_J$ ($J = 4, 5$)	600, 690
	$^3P_0 \to {}^3H_J$ ($J = 4$–6)	490, 545, 615, 640
	$^3P_0 \to {}^3F_J$ ($J = 2$–4)	700, 725
Nd	$^4F_{3/2} \to {}^4I_J$ ($J = 9/2$–$13/2$)	900, 1060, 1350
Sm	$^4G_{5/2} \to {}^6H_J$ ($J = 9/2$–$13/2$)	560, 595, 640, 700, 775
	$^4G_{5/2} \to {}^6F_J$ ($J = 1/2$–$9/2$)	870, 887, 926, 1010, 1150
Eu	$^5D_0 \to {}^7F_J$ ($J = 0$–6)	580, 590, 615, 650, 720, 750, 820
Gd	$^6P_{7/2} \to {}^8S_{7/2}$	315
Tb	$^5D_4 \to {}^7F_J$ ($J = 6$–0)	490, 540, 580, 620, 650, 660, 675
Dy	$^4F_{9/2} \to {}^6H_J$ ($J = 15/2$–$9/2$)	475, 570, 660, 750
	$^4I_{15/2} \to {}^6H_J$ ($J = 15/2$–$9/2$)	455, 540, 615, 695
Ho	$^5S_2 \to {}^5I_J$ ($J = 8, 7$)	545, 750
	$^7F_5 \to {}^5I_J$ ($J = 8, 7$)	650, 965
Er	$^4S_{3/2} \to {}^4I_J$ ($J = 15/2$–$13/2$)	545, 850
	$^4F_{9/2} \to {}^4I_{15/2}$	660
Tm	$^1D_2 \to {}^3F_4, {}^3H_4, {}^3F_J$ ($J = 3, 2$)	450, 650, 740, 775
	$^1G_4 \to {}^3H_6, {}^3F_4, {}^3H_5$	470, 650, 770
	$^3H_4 \to {}^3H_6$	800
Yb	$^2F_{5/2} \to {}^2F_{7/2}$	980

1.3.2 Promotion of 4f-4f Transitions

As already mentioned, 4f-4f transitions are intrinsically forbidden. First, these transitions accompany a change in spin, and therefore, they are spin forbidden. Second, since both the electric dipole operator and f orbitals have *ungerade* (odd) symmetry, electric dipole transitions are parity forbidden. However, the spin-orbit coupling invalidates the total spin quantum number S to relax the spin rule, and the effects of ligand field upon 4f-5d mixing also partially allow electric dipole transitions [52]. A mixture of opposite parities via crystal fields accompanied by partially allowed transitions was independently proposed by Brian Judd and George Offelt around the same time [53, 54]. The radiative rate constants (k_r) of these transitions are strongly affected by the geometrical symmetry around Ln(III) ions. The 4f-4f transition probabilities have been enhanced by introducing asymmetrical geometry around Ln(III) ions [55–58].

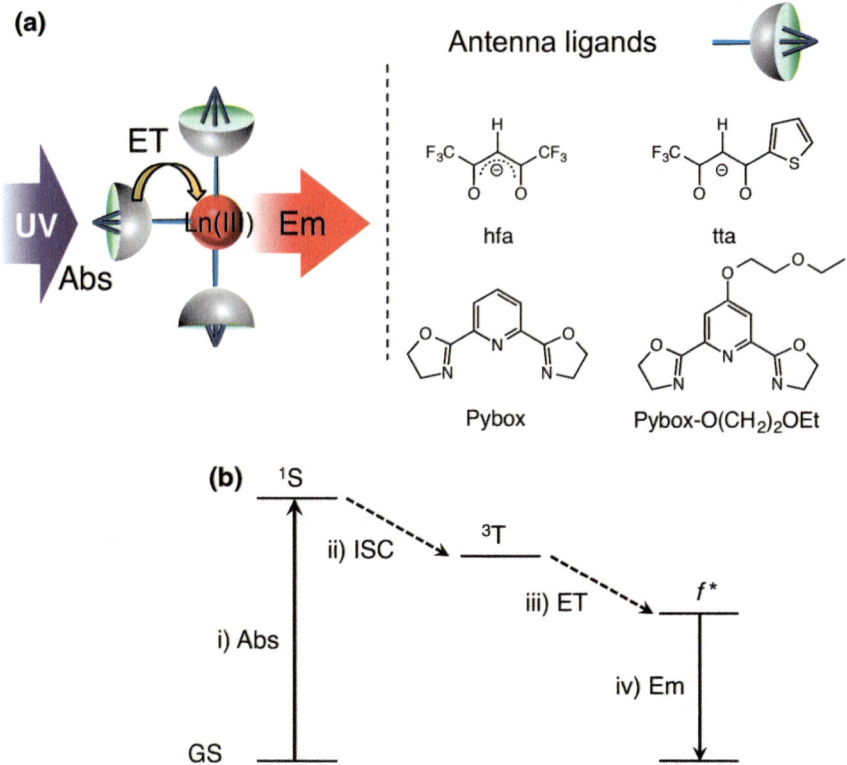

Fig. 1.3 **a** A schematic representation of antenna effect (left) and chemical structures of reported antenna ligands (right). **b** A simplified diagram of sensitized luminescence (Abs: absorption, ET: energy transfer, Em: emission, GS: ground state, ^{1}S: singlet excited state, ^{3}T: triplet excited state, ISC: intersystem crossing, f^*: emissive level of Ln(III) ions)

Hasegawa and co-workers reported the enhancement of luminescence by introducing an eight-coordinated square anti-prismatic (8-SAP: D_{4d}), a dodecahedral (8-TDH: D_{2d}), and a nine-coordinated monocapped square-antiprismatic (9-SAP: C_{4v}) molecular geometry (Fig. 1.4) [59–62]. In addition, seven-coordinated Ln(III) complexes with bulky tetramethylheptanedionate (tmh) ligands, including a monocapped octahedral (7-MCO: C_{3v}), a monocapped-trigonal-prismatic (7-MCTP: C_{2v}), and a pentagonal-bipyramidal (7-PBP: D_{5h}) structures, have recently been reported for further enhancement of 4f-4f transitions [63, 64]. These coordination geometry without inversion center effectively enhances radiative rates.

1.3 Enhancement of Luminescence Efficiency

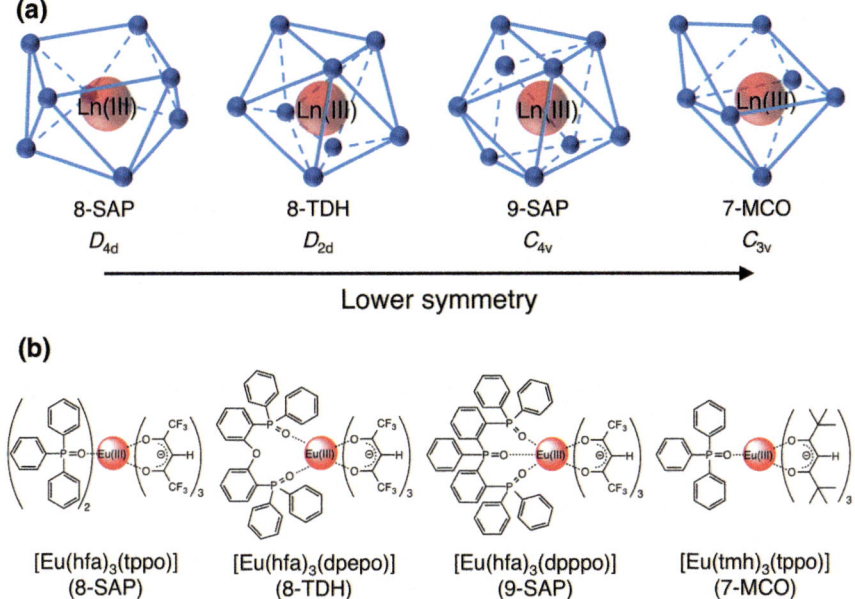

Fig. 1.4 a Ideal coordination geometrical structures of eight-coordinated square anti-prism (8-SAP), dodecahedron (8-TDH: D_{2d}), nine-coordinate monocapped square-antiprism (9-SAP: C_{4v}), and seven-coordinated monocapped octahedron (7-MCO: C_{3v}). b Chemical structures of reported asymmetrical Eu(III) coordination compounds [59–64]

1.3.3 Suppression of Vibrational Relaxation

Suppression of the non-radiative process is also essential for strong emission since Ln(III)-centered emission is easily quenched through vibrational relaxation. When the energy gaps between emissive and ground states of Ln(III) ions correspond well to the vibration of chemical bonds with higher vibrational quantum numbers ($v > 1$), Ln(III)-centered emission is non-radiatively quenched. Thus, high vibrational frequency bonds such as O–H (3600 cm^{-1}), N–H (3300 cm^{-1}), and C–H (2900 cm^{-1}) are known to be quenchers, and Ln(III) complexes with these chemical bonds generally exhibit large non-radiative rate constant (k_{nr}). Instead of these bonds, C=O (1650 cm^{-1}), C–F (1200 cm^{-1}), P=O (1120 cm^{-1}), O–D (2200 cm^{-1}), or C–D (2100 cm^{-1}) bonds can suppress the non-radiative relaxation, since higher vibrational quanta are needed and the process becomes inefficient (Fig. 1.5a).

Based on this idea, Hasegawa and co-workers first succeeded in observing Nd(III)-centered luminescence in an organic solvent [65]. In their work, deuterated hexafluoroacetylacetonato (hfa-D) ligands played an important role in providing both the antenna effect and the elimination of high-vibrational C–H and O–H bonds (Fig. 1.5b, left). Pikramenou and co-workers demonstrated long emission lifetimes of Yb(III) coordination compounds with fully fluorinated imidodiphosphinate

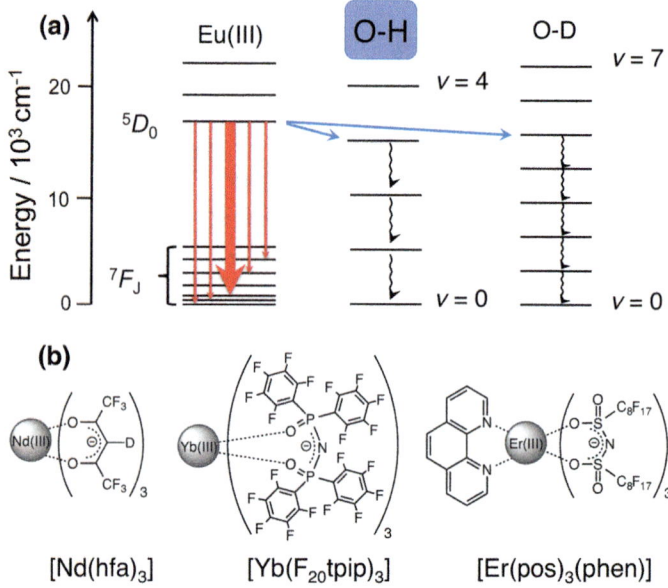

Fig. 1.5 **a** Radiative and non-radiative vibrational quenching process focused on Eu(III) ions. **b** Chemical structures of reported low-vibrational Ln(III) coordination compounds [64–66]

ligands (*N*-(*P,P*-di(pentafluorophinoyl))-*P,P*-dipentafluorophenylphosphinimidic acid: HF$_{20}$tpip), which was achieved by the absence of high energy X-H vibrations (Fig. 1.5b, center) [66]. Binnemans and co-workers also reported strong NIR luminescence of Er(III) complexes consisting of Er(III) ions, bis(perfluoroalkylsulfonyl)imides, and 1,10-phenanthroline [Er(pos)$_3$(phen)] (Fig. 1.5b, right) [67]. Suppression of molecular vibrations leads smaller non-radiative rate and stronger luminescence.

1.4 Assembled Structures of Coordination Compounds

In addition to the molecular design around metal ions, control of the entire morphology of coordination compounds is also required to explore their potential as functional materials. Precisely controlled assembly of multi-metal ions and organic bridging ligands provides novel physical or chemical properties that do not appear in individual components. The controlled assembly has recently become a common idea, particularly for transition metal coordination systems [68–72].

In 1990, Fujita and co-workers reported the formation of planar-squared Pd(II) coordination compounds [(en)Pd(4,4'-bpy)]$_4$(NO$_3$)$_8$ (en: ethylenediamine, bpy: bipyridine) by mixing Pd(II) ions and organic ligands at room temperature through a self-assembly mechanism (Fig. 1.6) [73]. The Pd(II) complex also has the ability

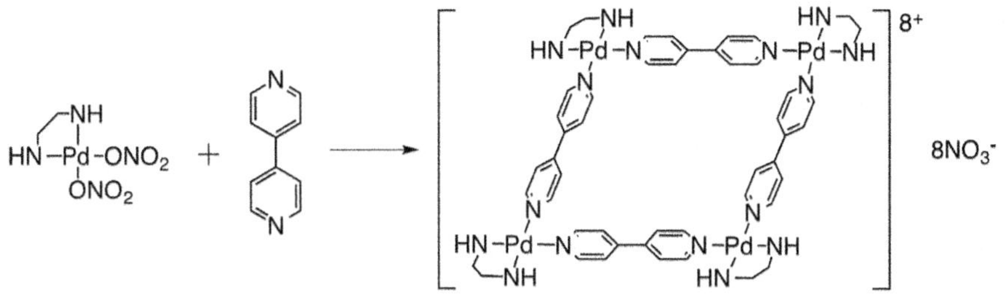

1.4 Assembled Structures of Coordination Compounds

Fig. 1.6 A synthetic scheme of a planar-squared Pd(II) complex [72]

to recognize an organic molecule in an aqueous medium. They have continuously reported the controlled assembly of polyhedral clusters using Pd(II), Co(II), Zn(II), and pyridine-based bridging ligands [74]. These results suggested that finite and single substrate was formed via self-assembly.

Assembled coordination polymers are also known to form characteristic 1-D chains, 2-D sheets, and 3-D networks depending on the moiety of organic bridging ligands [75, 76]. These structural features can provide additional functions such as electrical conductivity, redox activity, and specific ion/compound sensitivity as well as photophysical properties [77–82]. Niu and co-workers reported temperature-dependent magnetic susceptibility and photoluminescence of 2-D organic-inorganic architectures composed of Ln(III)-Cu(II) heterometallic phosphotungstates [83]. Wang developed Mg(II), Ni(II), and Co(II)-based 3-D networks linked by dibenzoic acid derivatives for highly sensitive and selective detection of heterocyclic compounds [84]. Vittal and co-workers also reported that 2-D sheets of emissive Zn(II) coordination compounds were assembled to 3-D networks via [2+2] cycloaddition reaction of bridging ligands [85].

These studies have exploited the designs of organic bridging ligands for construction and modification of a wide variety of multi-nuclear complexes, coordination polymers, and metal organic frameworks (MOFs). Thus, much larger, complicated, and controlled assemblies with advanced properties can be expected with appropriate designs of molecules.

1.5 Assembled Ln(III) Coordination Compounds

Ln(III)-based coordination polymers or MOFs have been regarded as important sources of practical luminescent materials due to their rigid scaffolds with versatile photophysical properties, and extensively studied in the recent two decades [86]. Carlos and co-workers reported a series of luminescent Ln(III)-based MOFs that are sensitive to anions, cations, and specific organic molecules [87]. Reddy reported thermally stable Eu(III) coordination polymers [88]. Hasegawa and co-workers also succeeded in providing thermal stability (>300 °C) and high luminescence efficiency

(70%) [89]. Numerous studies on Ln(III)-mixed MOFs have recently been carried out due to the stable, tunable, and ratiometric luminescence properties [90–94]. This class of luminescent materials has the advantage of being composed of thermodynamically stable backbones; however, the rigid frameworks (typically composed of Ln(III) ions and carboxylates) are usually unsuitable to accommodate bulky compounds, and allow adsorption of small organic solvents instead of ideal antenna ligands such as β-diketonates. Therefore, the design of assembled Ln(III) coordination compounds without inhibiting strong luminescence properties is highly desired.

The development of assembled Ln(III) coordination systems has relatively short history and been in the infancy compared to those of transition metal compounds. Since an integrated approach to molecular designs for strong luminescence with novel functions are still in progress, there is much room for improvement of photophysical, thermal, and mechanical properties by controlling assembled structures.

1.6 Objectives

An appropriate design of organic bridging ligands makes it possible to control the entire structures of assembled Ln(III) coordination compounds, which lead to novel photophysical, thermal, and mechanical properties.

For this thesis, control of the morphology of assembled Ln(III) coordination compounds was studied in order to achieve enhanced luminescence efficiency and further provide advanced physical properties such as glass-formability, temperature- and shock-sensitivity. The author achieved (i) high energy transfer efficiencies in dense-packed structures, (ii) tunable temperature sensitivity using Ln(III) mixed systems, (iii) glass-forming ability with hard-to-crystallize structures, and (iv) shock-sensitivity by introducing a specific face-to-face arrangement of substituents between molecules.

1.7 Contents of This Thesis

This thesis consists of seven chapters.

In this chapter, brief history of luminescent materials, previous studies on Ln(III) complexes and assembled coordination systems are reviewed, and the purpose of this study is described.

In Chap. 2, a packing system named "coordination zippers" with thiophene-based bridging ligands is proposed. The multiple intermolecular hydrogen bonds promote well-zipped and dense-packed coordination structures, resulting in dramatically improved thermal stability and photoluminescence (PL) efficiency via ligand excitation.

1.7 Contents of This Thesis

In Chap. 3, the control of photophysical properties in Tb(III)/Eu(III) mixed coordination polymers for tunable temperature sensitivity is described. The energy transfer efficiency between Ln(III) ions were studied using emission lifetime measurements.

In Chap. 4, the strategy for construction of luminescent lanthanide coordination compounds with glass-formability, "lanthanide coordination glasses," is described. Bent-angled bridging ligands with ethynyl groups are introduced to suppress crystal packing.

Based on the strategy suggested in Chaps. 3 and 4, extension of the molecular designs of coordination glasses and their temperature-responsive luminescence in a glassy state are demonstrated in Chap. 5.

In Chap. 6, the design of coordination polymers with triboluminescence (TL) properties is described. Strong TL materials were fabricated by inducing an intermolecular face-to-face arrangement of substituents. A significant TL/PL spectral difference was observed. The mechanisms of these phenomena are discussed on the basis of observations of Tb(III)/Eu(III) mixed coordination polymers.

Finally, in Chap. 7, summary and outlook of these studies are given.

References

1. E.F. Schubert, J.K. Kim, Science **308**, 1274–1278 (2005)
2. T.P. Yoon, M.A. Ischay, J.N. Du, Nat. Chem. **2**, 527–532 (2010)
3. F. Bonaccorso, Z. Sun, T. Hasan, A.C. Ferrari, Nat. Photonics **4**, 611–622 (2010)
4. N. Holonyak, S.F. Bevacqua, Appl. Phys. Lett. **1**, 82–83 (1962)
5. S. Nakamura, J. Cryst. Growth **145**, 911–917 (1994)
6. S. Nakamura, T. Mukai, M. Senoh, Appl. Phys. Lett. **64**, 1687–1689 (1994)
7. H. Amano, M. Kito, K. Hiramatsu, I. Akasaki, Jpn. J. Appl. Phys. **28**, L2112–L2114 (1989)
8. I. Akasaki, H. Amano, M. Kito, K. Hiramatsu, J. Lumin. **48–49**, 666–670 (1991)
9. C. Adachi, M.A. Baldo, M.E. Thompson, S.R. Forrest, J. Appl. Phys. **90**, 5048–5051 (2001)
10. J. Kido, H. Hayase, K. Hongawa, K. Nagai, K. Okuyama, Appl. Phys. Lett. **65**, 2124–2126 (1994)
11. J. Kido, K. Nagai, Y. Ohashi, Chem. Lett. **19**, 657–660 (1990)
12. C.D. Muller, A. Falcou, N. Reckefuss, M. Rojahn, V. Wiederhirn, P. Rudati, H. Frohne, O. Nuyken, H. Becker, K. Meerholz, Nature **421**, 829–833 (2003)
13. K. Kuriki, Y. Koike, Y. Okamoto, Chem. Rev. **102**, 2347–2356 (2002)
14. J.M. Costa-Fernández, R. Pereiro, A. Sanz-Medel, Trends Anal. Chem. **25**, 207–218 (2006)
15. M.A. Baldo, D.F. O'Brien, Y. You, A. Shoustikov, S. Sibley, M.E. Thompson, S.R. Forrest, Nature **395**, 151–154 (1998)
16. J.M. Phillips, M.E. Coltrin, M.H. Crawford, A.J. Fischer, M.R. Krames, R. Mueller-Mach, G.O. Mueller, Y. Ohno, L.E.S. Rohwer, J.A. Simmons, J.Y. Tsao, Laser Photonics Rev. **1**, 307–333 (2007)
17. P. Pust, V. Weiler, C. Hecht, A. Tücks, A.S. Wochnik, A.-K. Henß, D. Wiechert, C. Scheu, P.J. Schmidt, W. Schnick, Nat. Mater. **13**, 891–896 (2014)
18. C. Joblin, F. Salama, L. Allamandola, J. Chem. Phys. **110**, 7287–7297 (1999)
19. J. Donovalova, M. Cigan, H. Stankovicova, J. Gaspar, M. Danko, A. Gaplovsky, P. Hrdlovic, Molecules **17**, 3259–3276 (2012)
20. K. Hara, Z.S. Wang, T. Sato, A. Furube, R. Katoh, H. Sugihara, Y. Dan-Oh, C. Kasada, A. Shinpo, S. Suga, J. Phys. Chem. B **109**, 15476–15482 (2005)
21. M.-C. Choi, Y. Kim, C.-S. Ha, Prog. Polym. Sci. **33**, 581–630 (2008)

22. M.H. Park, T.H. Han, Y.H. Kim, S.H. Jeong, Y. Lee, H.K. Seo, H. Cho, T.W. Lee, J. Photonics Energy **5** (2015)
23. T.H. Han, Y. Lee, M.R. Choi, S.H. Woo, S.H. Bae, B.H. Hong, J.H. Ahn, T.W. Lee, Nat. Photonics **6**, 105–110 (2012)
24. M.D. Ward, P.R. Raithby, Chem. Soc. Rev. **42**, 1619–1636 (2013)
25. S.V. Eliseeva, J.C.G. Bünzli, Chem. Soc. Rev. **39**, 189–227 (2010)
26. J.-C.G. Bünzli, C. Piguet, Chem. Soc. Rev. **34**, 1048–1077 (2005)
27. K. Binnemans, Chem. Rev. **109**, 4283–4374 (2009)
28. S.J. Butler, D. Parker, Chem. Soc. Rev. **42**, 1652–1666 (2013)
29. S. Petoud, G. Muller, E.G. Moore, J. Xu, J. Sokolnicki, J.P. Riehl, U.N. Le, S.M. Cohen, K.N. Raymond, J. Am. Chem. Soc. **129**, 77–83 (2007)
30. S. Petoud, S.M. Cohen, J.-C.G. Bünzli, K.N. Raymond, J. Am. Chem. Soc. **125**, 13324–13325 (2003)
31. J.-C.G. Bünzli, S. Comby, A.S. Chauvin, C.D.B. Vandevyver, J. Rare Earths **25**, 257–274 (2007)
32. J.-C.G. Bünzli, Chem. Rev. **110**, 2729–2755 (2010)
33. A. de Bettencourt-Dias, Introduction to lanthanide ion luminescence, in *Luminescence of Lanthanide Ions in Coordination Compounds and Nanomaterials* (Wiley, 2014), pp. 1–48
34. A. de Bettencourt-Dias, Dalton Trans. 2229–2241 (2007)
35. S.V. Eliseeva, M. Ryazanov, F. Gumy, S.I. Troyanov, L.S. Lepnev, J.-C.G. Bünzli, N.P. Kuzmina, Eur. J. Inorg. Chem. 4809–4820 (2006)
36. Y. Ohishi, T. Kanamori, T. Kitagawa, S. Takahashi, E. Snitzer, G.H. Sigel, Opt. Lett. **16**, 1747–1749 (1991)
37. L.H. Slooff, A. Polman, M.P.O. Wolbers, F. van Veggel, D.N. Reinhoudt, J.W. Hofstraat, J. Appl. Phys. **83**, 497–503 (1998)
38. L. Armelao, S. Quici, F. Barigelletti, G. Accorsi, G. Bottaro, M. Cavazzini, E. Tondello, Coord. Chem. Rev. **254**, 487–505 (2010)
39. S.I. Weissman, J. Chem. Phys. **10**, 214–217 (1942)
40. G.A. Crosby, Mol. Cryst. **1**, 37–81 (1966)
41. G.A. Crosby, R.E. Whan, J. Chem. Phys. **36**, 863–865 (1962)
42. G.A. Crosby, R.E. Whan, R.M. Alire, J. Chem. Phys. **34**, 743–748 (1961)
43. K. Binnemans, Chapter 225—Rare-earth beta-diketonates, in *Handbook on the Physics and Chemistry of Rare Earths*, ed. by J.-C.G.Bünzli, K.A. Gschneidner, K.P. Vitalij (Elsevier, 2005), pp. 107–272
44. A. de Bettencourt-Dias, S. Bauer, S. Viswanathan, B.C. Maull, A.M. Ako, Dalton Trans. **41**, 11212–11218 (2012)
45. E.S. Andreiadis, N. Gauthier, D. Imbert, R. Demadrille, J. Pécaut, M. Mazzanti, Inorg. Chem. **52**, 14382–14390 (2013)
46. S.V. Eliseeva, D.N. Pleshkov, K.A. Lyssenko, L.S. Lepnev, J.C.G. Bünzli, N.P. Kuzminat, Inorg. Chem. **49**, 9300–9311 (2010)
47. N.B.D. Lima, S.M.C. Goncalves, S.A. Junior, A.M. Simas, Sci. Rep. **3** (2013)
48. S.V. Eliseeva, O.V. Kotova, F. Gumy, S.N. Semenov, V.G. Kessler, L.S. Lepnev, J.-C.G. Bünzli, N.P. Kuzmina, J. Phys. Chem. A **112**, 3614–3626 (2008)
49. A. de Bettencourt-Dias, P.S. Barber, S. Viswanathan, D.T. de Lill, A. Rollett, G. Ling, S. Altun, Inorg. Chem. **49**, 8848–8861 (2010)
50. A. de Bettencourt-Dias, P.S. Barber, S. Bauer, J. Am. Chem. Soc. **134**, 6987–6994 (2012)
51. A. de Bettencourt-Dias, S. Viswanathan, A. Rollett, J. Am. Chem. Soc. **129**, 15436–15437 (2007)
52. B.G. Wybourne, Mol. Phys. **101**, 899–901 (2003)
53. B.R. Judd, Phys. Rev. **127**, 750–761 (1962)
54. G.S. Ofelt, J. Chem. Phys. **37**, 511–520 (1962)
55. S.F. Mason, R.D. Peacock, B. Stewart, Chem. Phys. Lett. **29**, 149–153 (1974)
56. A.F. Kirby, F.S. Richardson, J. Phys. Chem. **87**, 2544–2556 (1983)

References

57. M. Montalti, L. Prodi, N. Zaccheroni, L. Charbonnière, L. Douce, R. Ziessel, J. Am. Chem. Soc. **123**, 12694–12695 (2001)
58. K. Driesen, P. Lenaerts, K. Binnemans, C. Gorller-Walrand, Phys. Chem. Chem. Phys. **4**, 552–555 (2002)
59. T. Harada, Y. Nakano, M. Fujiki, M. Naito, T. Kawai, Y. Hasegawa, Inorg. Chem. **48**, 11242–11250 (2009)
60. T. Harada, H. Tsumatori, K. Nishiyama, J. Yuasa, Y. Hasegawa, T. Kawai, Inorg. Chem. **51**, 6476–6485 (2012)
61. K. Miyata, T. Nakagawa, R. Kawakami, Y. Kita, K. Sugimoto, T. Nakashima, T. Harada, T. Kawai, Y. Hasegawa, Chem. Eur. J. **17**, 521–528 (2011)
62. K. Miyata, Y. Hasegawa, Y. Kuramochi, T. Nakagawa, T. Yokoo, T. Kawai, Eur. J. Inorg. Chem. 4777–4785 (2009)
63. K. Yanagisawa, T. Nakanishi, Y. Kitagawa, T. Seki, T. Akama, M. Kobayashi, T. Taketsugu, H. Ito, K. Fushimi, Y. Hasegawa, Eur. J. Inorg. Chem. 4769–4774 (2015)
64. K. Yanagisawa, Y. Kitagawa, T. Nakanishi, T. Akama, M. Kobayashi, T. Seki, K. Fushimi, H. Ito, T. Taketsugu, Y. Hasegawa, Eur. J. Inorg. Chem. 3843–3848 (2017)
65. Y. Hasegawa, T. Ohkubo, K. Sogabe, Y. Kawamura, Y. Wada, N. Nakashima, S. Yanagida, Angew. Chem. Int. Ed. **39**, 357–360 (2000)
66. P.B. Glover, A.P. Bassett, P. Nockemann, B.M. Kariuki, R. Van Deun, Z. Pikramenou, Chem. Eur. J. **13**, 6308–6320 (2007)
67. R. Van Deun, P. Nockemann, C. Gorller-Walrand, K. Binnemans, Chem. Phys. Lett. **397**, 447–450 (2004)
68. M. Burnworth, L.M. Tang, J.R. Kumpfer, A.J. Duncan, F.L. Beyer, G.L. Fiore, S.J. Rowan, C. Weder, Nature **472**, 334–U230 (2011)
69. H. Furukawa, K.E. Cordova, M. O'Keeffe, O.M. Yaghi, Science **341**, 974 (2013)
70. A.V. Zhukhovitskiy, M.Z. Zhong, E.G. Keeler, V.K. Michaelis, J.E.P. Sun, M.J.A. Hore, D.J. Pochan, R.G. Griffin, A.P. Willard, J.A. Johnson, Nat. Chem. **8**, 33–41 (2016)
71. T. Fukino, H. Joo, Y. Hisada, M. Obana, H. Yamagishi, T. Hikima, M. Takata, N. Fujita, T. Aida, Science **344**, 499–504 (2014)
72. A. Tsuda, Y. Nagamine, R. Watanabe, Y. Nagatani, N. Ishii, T. Aida, Nat. Chem. **2**, 977–983 (2010)
73. M. Fujita, J. Yazaki, K. Ogura, J. Am. Chem. Soc. **112**, 5645–5647 (1990)
74. Y. Inokuma, M. Kawano, M. Fujita, Nat. Chem. **3**, 349–358 (2011)
75. T.R. Cook, Y.R. Zheng, P.J. Stang, Chem. Rev. **113**, 734–777 (2013)
76. S. Kitagawa, R. Kitaura, S. Noro, Angew. Chem. Int. Ed. **43**, 2334–2375 (2004)
77. A. Gallego, O. Castillo, C.J. Gomez-Garcia, F. Zamora, S. Delgado, Inorg. Chem. **51**, 718–727 (2012)
78. M.I.J. Polson, E.A. Medlycott, G.S. Hanan, L. Mikelsons, N.L. Taylor, M. Watanabe, Y. Tanaka, F. Loiseau, R. Passalacqua, S. Campagna, Chem. Eur. J. **10**, 3640–3648 (2004)
79. F. Puntoriero, S. Campagna, A.M. Stadler, J.M. Lehn, Coord. Chem. Rev. **252**, 2480–2492 (2008)
80. A. Rana, S.K. Jana, T. Pal, H. Puschmann, E. Zangrando, S. Dalai, J. Solid State Chem. **216**, 49–55 (2014)
81. J.N. Hao, B. Yan, J. Mater. Chem. A **3**, 4788–4792 (2015)
82. L.N. Zhang, A.L. Liu, Y.X. Liu, J.X. Shen, C.X. Du, H.W. Hou, Inorg. Chem. Commun. **56**, 137–140 (2015)
83. S.S. Shang, J.W. Zhao, L.J. Chen, Y.Y. Li, J.L. Zhang, Y.Z. Li, J.Y. Niu, J. Solid State Chem. **196**, 29–39 (2012)
84. Y.X. Guo, X. Feng, T.Y. Han, S. Wang, Z.G. Lin, Y.P. Dong, B. Wang, J. Am. Chem. Soc. **136**, 15485–15488 (2014)
85. R. Medishetty, R. Tandiana, L.L. Koh, J. Vittal, Chem. Eur. J. **20**, 1231–1236 (2014)
86. Y. Hasegawa, T. Nakanishi, RSC Adv. **5**, 338–353 (2015)
87. J. Rocha, L.D. Carlos, F.A.A. Paz, D. Ananias, Chem. Soc. Rev. **40**, 926–940 (2011)
88. A.R. Ramya, D. Sharma, S. Natarajan, M.L.P. Reddy, Inorg. Chem. **51**, 8818–8826 (2012)

89. K. Miyata, T. Ohba, A. Kobayashi, M. Kato, T. Nakanishi, K. Fushimi, Y. Hasegawa, ChemPlusChem **77**, 277–280 (2012)
90. K.A. White, D.A. Chengelis, K.A. Gogick, J. Stehman, N.L. Rosi, S. Petoud, J. Am. Chem. Soc. **131**, 18069–18071 (2009)
91. J.Y. An, C.M. Shade, D.A. Chengelis-Czegan, S. Petoud, N.L. Rosi, J. Am. Chem. Soc. **133**, 1220–1223 (2011)
92. Y.J. Cui, H. Xu, Y.F. Yue, Z.Y. Guo, J.C. Yu, Z.X. Chen, J.K. Gao, Y. Yang, G.D. Qian, B.L. Chen, J. Am. Chem. Soc. **134**, 3979–3982 (2012)
93. X.T. Rao, T. Song, J.K. Gao, Y.J. Cui, Y. Yang, C.D. Wu, B.L. Chen, G.D. Qian, J. Am. Chem. Soc. **135**, 15559–15564 (2013)
94. Y.J. Cui, B.L. Chen, G.D. Qian, Coord. Chem. Rev. **273**, 76–86 (2014)

Chapter 2
Luminescent Lanthanide Coordination Zippers with Dense-Packed Structures for High Energy Transfer Efficiencies

Abstract Novel Eu(III) coordination polymers [Eu(hfa)$_3$(dpt)]$_n$ [dpt: 2,5-bis(diphenylphosphoryl)thiophene] and [Eu(hfa)$_3$(dpedot)]$_n$ [dpedot: 3,4-bis(diphenylphosphoryl)ethylenedioxythiophene] were designed for dense structures with high energy transfer efficiency. The zig-zag orientation of single polymer chains induced the formation of dense-packed coordination structures with multiple inter-molecular hydrogen bonds. These polymers exhibited high intrinsic emission quantum yields (~80%) due to their asymmetrical and low-vibrational coordination structures. The significant energy transfer efficiencies of up to 80% were also achieved.

Keywords Europium · Coordination polymer · Luminescence
Energy transfer

2.1 Introduction

Developing luminescent molecular materials with high quantum efficiencies is required for applications in light-emitting devices, bio-probes, chemical and physical sensors [1–7]. There have been many reports on luminescent organic and coordination compounds. Swager and co-workers developed amplifying fluorescent conjugated polymers for biological and chemical sensors such as highly explosive trinitrotoluene in seawater [8]. Adachi et al. reported very high phosphorescence efficiency of Ir(III) coordination compounds in organic light-emitting devices [9].

Among these materials, Ln(III) coordination compounds are promising candidates for pure and strong luminophores as described in Chap. 1 [10–17]. The overall emission quantum yields of these compounds are described as:

$$\Phi_{total} = \Phi_{ff} \times \eta_{sens} = \frac{k_r}{k_r + k_{nr}} \times \eta_{sens} \qquad (2.1)$$

© Springer Nature Singapore Pte Ltd. 2018
Y. Hirai, *Assembled Lanthanide Complexes with Advanced Photophysical Properties*,
Springer Theses, https://doi.org/10.1007/978-981-10-8932-9_2

where Φ_{ff} is Ln(III)-centered emission quantum yields, η_{sens} is efficiency of the sensitization process, and k_r and k_{nr} are radiative and non-radiative rate constants, respectively.

Ln(III) coordination compounds with high Φ_{ff} have been successfully synthesized by introducing asymmetrical and low-vibrational coordination structures, resulting in large k_r and small k_{nr} values [18, 19] In order to advance these compounds for industrial use, Ln(III) coordination polymers with rigid multi-dimensional networks have been developed over the past two decades [20–22]. The reported Ln(III) coordination polymers exhibit high Φ_{ff} (>70%) and thermal stability (>200 °C); however, their energy transfer efficiencies from antenna ligands to Ln(III) ions (η_{sens}) are estimated to be approximately 50% [23, 24]. Ln(III) coordination polymers with high η_{sens} are required for ideal optical devices.

The author here focused on the intra-ligand charge transfer (ILCT) states via charge redistribution of hfa ligands in Ln(III) coordination polymers to enhance the energy transfer efficiency. The ILCT states are formed under specific packing structures and found to contribute to the photosensitization process in Ln(III) complexes [25]. Eliseeva and co-workers recently reported the considerable enhancement of ligand-to-metal energy transfer efficiency for both VIS and NIR emissive Ln(III) coordination compounds under existence of low-lying ILCT states [26, 27]. Hasegawa and co-workers have also demonstrated efficient photosensitized luminescence of Eu(III) coordination polymers with ILCT bands [28]. The formation of ILCT states should affect the efficiency of energy transfer from ligands to Ln(III) ions. Since these ligand-based electronic effects are characteristic in solid state, dense and tight coordination structures in crystal units are expected to induce the formation of ILCT states, leading to high Φ_{tot}. The author here considers that luminescent Ln(III) coordination polymers with specific ILCT states can be constructed by zig-zag chained polymers, "coordination zippers."

In this chapter, the novel thiophene-based bridging ligands, 2,5-bis(diphenylphosphoryl)thiophene (dpt), 3,4-bis(diphenylphosphoryl)ethylenedioxithiophene (dpedot), and corresponding Eu(III) coordination polymers ([Eu(hfa)$_3$(dpt)]$_n$ and [Eu(hfa)$_3$(dpedot)]$_n$, respectively, Fig. 2.1a) were prepared to form zig-zag polymer chains with close-packed structures in solid state. Small aromatic core and bent angles of thiophene-based bridges are expected to reduce the distances between Eu(hfa)$_3$ units in single polymer chains. The zig-zag polymer moiety is also expected to induce interdigitation between polymer chains to form intermolecular CH/F and CH/π hydrogen bonds through bridging ligands and antenna ligands. The Eu(III) coordination polymer with linear-typed bridges ([Eu(hfa)$_3$(dpbp)]$_n$ [24], Fig. 2.1b) was also prepared for comparison. Their photophysical properties were estimated by UV/Vis absorption, emission, and excitation spectra. Thermal stability is discussed on the basis of the results of thermogravimetric analyses and single crystal X-ray diffraction.

2.2 Experimental Section

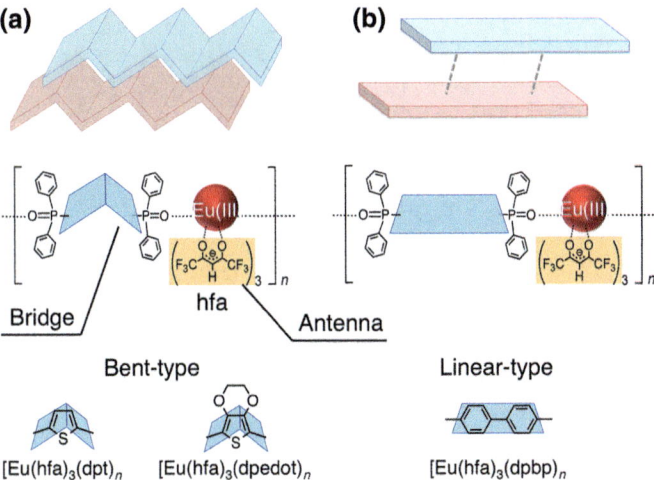

Fig. 2.1 Diagram of **a** Eu(III) coordination zippers with bent-type bridges and **b** Eu(III) coordination polymers with linear-type bridges

2.2 Experimental Section

2.2.1 General

Europium acetate *n*-hydrate (99.9%), *n*-BuLi (in *n*-hexane, 1.6 M), and hydrogen peroxide were purchased from Kanto Chemical Co., Inc. 2,5-Dibromothiophene, 3,4-ethylenedioxythiophene, and chlorodiphenylphosphine (PPh$_2$Cl) were obtained from Tokyo Chemical Industry Co., Ltd. All other chemicals and solvents were reagent grade and were used without further purification.

2.2.2 Apparatus

^{1}H NMR (400 MHz) spectra were recorded on a JEOL ECS400. Chemical shifts were reported in δ ppm, referenced to an internal tetramethylsilane standard for ^{1}H NMR spectroscopy. Infrared spectra were recorded on a JASCO FTIR-420 spectrometer using KBr pellets. Elemental analyses were performed by an Exeter Analytical CE440. Mass spectrometry was performed by a Thermo Scientific Exactive (ESI-MS) and a JEOL JMS-700TZ (FAB-MS). Thermogravimetric analysis (TGA) was performed on a Seiko Instruments Inc. EXSTAR 6000 (TG-DTA 6300) in an argon atmosphere at a heating rate of 5 °C min^{-1}.

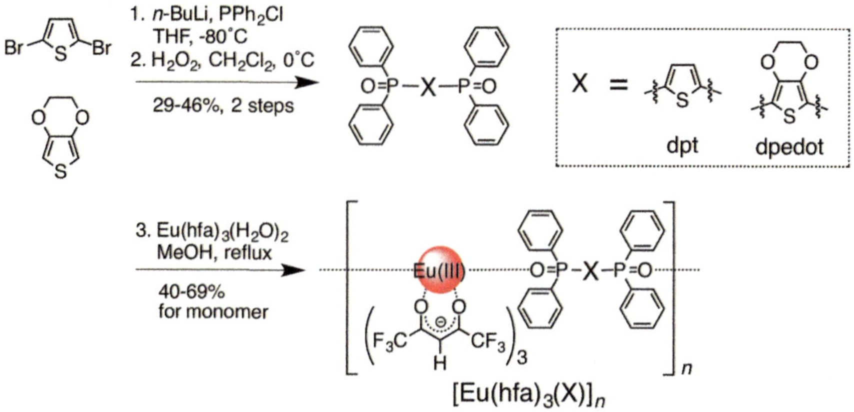

Scheme 2.1 Synthetic schemes of phosphine oxide ligands and Eu(III) coordination polymers

2.2.3 Syntheses

The organic bridging ligands and corresponding coordination compounds were synthesized following Scheme 2.1.

Preparation of [Eu(hfa)$_3$(H$_2$O)]$_2$:

Europium acetate hydrate (2.0 g, 5.8 mmol) was dissolved in distilled water (30 mL). Hexafluoroacetylacetone (4.0 g, 19 mmol) was added dropwise to the solution and let stirred for 3 h at room temperature to form pale yellow precipitates. The reaction mixture was filtered and washed with distilled water and chloroform for several times. The resulting powder was used without further purification for the next step.

Yield 4.1 g (89%). IR (KBr): 1650 (st, C=O), 1258-1145 (st, C-F) cm^{-1}. Anal. Calcd for C$_{15}$H$_7$EuF$_{18}$O$_8$: C, 22.27; H, 0.87%. Found: C, 22.12; H, 1.01%.

Preparation of 2,5-bis(diphenylphosphoryl)thiophene (dpt):

In a degassed 3-neck round-bottomed flask (300 mL vol.), 2,5-dibromothiophene (2 mL, 17.7 mmol) was dissolved in dry THF (70 mL) under argon atmosphere, then stirred until a homogeneous solution was formed at room temperature. A solution of n-BuLi (28 mL, 44 mmol) was added dropwise to the solution at −80 °C. The addition was completed in ca. 15 min. The mixture was allowed to stir for 3 h, after which a PPh$_2$Cl (8.0 mL, 45 mmol) was added dropwise at −80 °C, then the solution became cloudy. The mixture was gradually brought to room temperature, and stirred for 14 h to form clouded yellow solution. The product was extracted with dichloromethane and dried over anhydrous MgSO$_4$. The solvent was concentrated and dissolved in dichloromethane (50 mL) in a flask. The solution was cooled to 0 °C and then 30% H$_2$O$_2$ aqueous solution (24 mL) was added to it. The reaction mixture was stirred for 2 h. The product was extracted with dichloromethane and

2.2 Experimental Section

the obtained crude powder was washed with ethyl acetate for several times to afford white powder.

Yield: 2.3 g (29%). ^{1}H NMR (400 MHz, CDCl$_3$, 25 °C) δ 7.79–7.86 (m, 8H, –CH), δ 7.64–7.67 (m, 2H, –CH), δ 7.43–7.49 (m, 4H, –CH), δ 7.35–7.41 (m, 8H, –CH) ppm. ESI-Mass (m/z): calcd for C$_{28}$H$_{23}$O$_2$P$_2$S [M+H]$^+$, 485.09; found, 485.09. Anal. Calcd for C$_{28}$H$_{22}$O$_2$P$_2$S: C, 69.41; H, 4.58%. Found: C, 68.90; H, 4.57%.

Preparation of 3,4-bis(diphenylphosphoryl)ethylenedioxythiophene (dpedot):

In a degassed 3-neck round-bottomed flask (300 mL vol.), 3,4-ethylenedioxythiophene (2 mL, 18.5 mmol) was dissolved in dry THF (80 mL) under argon atmosphere, then stirred until a homogeneous solution was formed at room temperature. A solution of n-BuLi (29 mL, 46 mmol) was added dropwise to the solution at −80 °C. The addition was completed in ca. 15 min. The mixture was allowed to stir for 2 h, after which a PPh$_2$Cl (8.5 mL, 46 mmol) was added dropwise at −80 °C. The mixture was gradually brought to room temperature, and stirred for 4 h. The product was extracted with dichloromethane and dried over anhydrous MgSO$_4$. The solvent was concentrated and dissolved in dichloromethane (50 mL) in a flask. The solution was cooled to 0 °C and then 30% H$_2$O$_2$ aqueous solution (25 mL) was added to it. The reaction mixture was stirred for 2 h. The product was extracted with dichloromethane and the obtained crude powder was washed with ethyl acetate for several times to afford pale brown powder.

Yield: 4.6 g (46%). ^{1}H NMR (400 MHz, CDCl$_3$, 25 °C) δ 7.68–7.74 (m, 8H, –CH), δ 7.51–7.55 (m, 4H, –CH), δ 7.42–7.46 (m, 8H, –CH), δ 4.12 (s, 4H, –CH$_2$) ppm. ESI-Mass (m/z): calcd for C$_{30}$H$_{25}$O$_4$P$_2$S [M+H]$^+$, 543.09; found, 543.09. Anal. Calcd for C$_{30}$H$_{24}$O$_4$P$_2$S: C, 66.42; H, 4.46%. Found: C, 66.04; H, 4.53%.

Preparation of [Eu(hfa)$_3$(dpt)]$_n$:

Phosphine oxide ligand (0.39 g, 0.80 mmol) and Eu(hfa)$_3$(H$_2$O)$_2$ (0.64 g, 0.80 mmol) were dissolved in methanol (40 mL), respectively. The solutions were mixed and refluxed for 3 h to form white precipitates. The obtained white solid was washed with methanol.

Yield: 0.40 g (40% for monomer). IR (KBr): 1655 (st, C=O), 1143 (st, P=O) cm^{-1}. FAB-Mass (m/z): calcd for C$_{38}$H$_{24}$EuF$_{12}$O$_6$P$_2$S [M-hfa]$^+$: 1051.0, found, 1050.8. Anal. Calcd for C$_{43}$H$_{25}$EuF$_{18}$O$_8$P$_2$S: C, 41.07; H, 2.00%. Found: C, 40.91; H, 2.12%.

Preparation of [Eu(hfa)$_3$(dpedot)]$_n$:

Phosphine oxide ligand (0.43 g, 0.80 mmol) and Eu(hfa)$_3$(H$_2$O)$_2$ (0.64 g, 0.80 mmol) were dissolved in methanol, respectively. The solutions were mixed and refluxed for 3 h. The reaction mixture was concentrated and washed with chloroform. The solvent was evaporated and re-dissolved in 50 °C methanol for recrystallization. The obtained crystals were washed with −20 °C methanol.

Yield: 0.73 g (69% for monomer). IR (KBr): 1657 (st, C=O), 1145 (st, P=O) cm^{-1}. FAB-Mass (m/z): calcd for C$_{40}$H$_{26}$EuF$_{12}$O$_8$P$_2$S [M-hfa]$^+$: 1109.0, found, 1109.0. Anal. Calcd for C$_{45}$H$_{27}$EuF$_{18}$O$_{10}$P$_2$S: C, 41.08; H, 2.07%. Found: C, 40.92; H, 2.37%.

2.2.4 Crystallography

Single crystals of [Eu(hfa)$_3$(dpt)]$_n$ were prepared by liquid-liquid diffusion of dpt/dichloromethane solution and Eu(hfa)$_3$(H$_2$O)$_2$/methanol solution. A colorless needle-like crystal of [Eu(hfa)$_3$(dpt)]$_n$ was mounted on a MicroLoop 75 μm using Paratone-N. The measurement was performed on a Rigaku XtaLAB P200 diffractometer using multi-layer mirror monochromated Mo-K$_\alpha$ radiation. Single crystals of [Eu(hfa)$_3$(dpedot)]$_n$ were prepared by slow evaporation of [Eu(hfa)$_3$(dpedot)]$_n$/methanol solution. A colorless block-shaped single crystal of [Eu(hfa)$_3$(dpedot)]$_n$ was mounted on a MiTiGen micromesh using Paratone-N. The measurement was performed on a Rigaku R-AXIS RAPID diffractometer using graphite monochromated Mo-K$_\alpha$ radiation. Non-hydrogen atoms were refined anisotropically. Hydrogen atoms were refined using the riding model. All calculations were performed using the CrystalStructure crystallographic software package. CIF data was confirmed by using the checkCIF/PLATON service.

2.2.5 Assignment of Coordination Geometry

To describe the geometry of Ln(III) complexes and to evaluate the degree of distortion from ideal geometry, the "shape measure" criterion S [29], suggested by Raymond and co-workers, is estimated using the following method (Fig. 2.2) and Eq. (2.2) on the basis of crystal data.

$$S = \min \sqrt{\left(\frac{1}{m}\right) \sum_{i=1}^{m} (\delta_i - \theta_i)^2} \qquad (2.2)$$

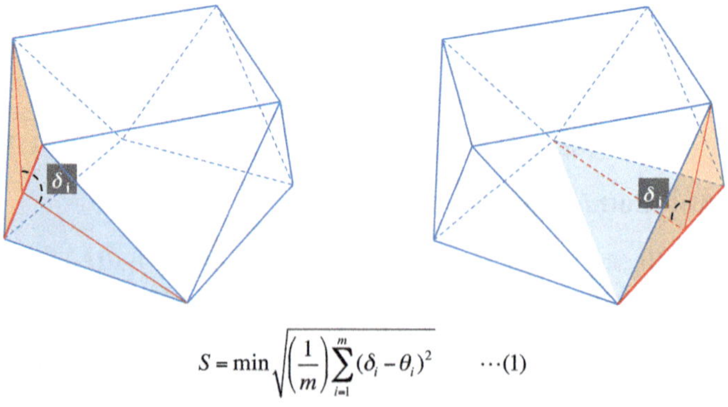

Fig. 2.2 Dihedral angles (δ_i) in an 8-SAP structure

1. Ln(III) ions and O atoms in the first coordination sphere are connected by lines.
2. The observed dihedral angle δ_i (angle between the normal and adjacent faces) along the ith edge and θ_i (the same angle for the corresponding ideal polyhedral shape) are measured.
3. The squared difference of δ_i and θ_i is estimated.
4. 2 and 3 are executed for all edges, and the sum of squared differences is obtained by dividing the number of edges ($m = 18$ in this study).
5. The square root of 4 is calculated.
6. Procedure 2–5 are executed for all possible geometries, and the smallest value is defined as S [deg] for the Ln(III) complex.

2.2.6 Optical Measurements

UV-Vis absorption spectra were recorded on a JASCO V-670 spectrometer. Emission and excitation spectra were recorded on a HORIBA Fluorolog-3 spectrofluorometer and corrected for the response of the detector system. Emission lifetimes (τ_{obs}) were measured using the third harmonics (355 nm) of a Q-switched Nd:YAG laser (Spectra Physiscs, INDI-50, fwhm = 5 ns, λ = 1064 nm) and a photomultiplier (Hamamatsu photonics, R5108, response time ≤1.1 ns). The Nd:YAG laser response was monitored with a digital oscilloscope (Sony Tektronix, TDS3052, 500 MHz) synchronized to the single-pulse excitation. Emission lifetimes were determined from the slope of logarithmic plots of the decay profiles. The emission quantum yields excited at 380 nm (Φ_{tot}) were estimated using JASCO F-6300-H spectrometer attached with JASCO ILF-533 integrating sphere unit ($\varphi = 100$ mm). The wavelength dependence of the detector response and the beam intensity of Xe light source for each spectrum were calibrated using a standard light source.

2.2.7 Estimation of Intrinsic Emission Quantum Yields

The photophysical parameters such as intrinsic emission quantum yields (Φ_{ff}), radiative (k_r) and non-radiative (k_{nr}) rate constants can be related in the simplified Eqs. (2.3)–(2.5) in the case of Eu(III) complexes due to the character of purely magnetic dipole transition $^5D_0 \rightarrow {}^7F_1$ [19, 30].

$$\Phi_{ff} = \frac{k_r}{k_r + k_{nr}} = \frac{\tau_{obs}}{\tau_{rad}} \quad (2.3)$$

$$k_r = \frac{1}{\tau_{rad}} = A_{MD,0} n^3 \left(\frac{I_{tot}}{I_{MD}}\right) \quad (2.4)$$

$$k_{nr} = \frac{1}{\tau_{obs}} - \frac{1}{\tau_{rad}} \quad (2.5)$$

Table 2.1 Photophysical parameters of Eu(III) coordination polymers [Eu(hfa)$_3$(X)]$_n$ in solid state

Sample	Φ_{tot}[a] /%	Φ_{ff}[b] /%	η_{sens}[a] /%	τ_{obs}[c] /ms	k_r[b] /s^{-1}	k_{nr}[b] /s^{-1}
[Eu(hfa)$_3$(dpt)]$_n$	60	75	80	0.75	1.0×10^3	3.3×10^2
[Eu(hfa)$_3$(dpedot)]$_n$	56	85	66	0.93	9.1×10^2	1.6×10^2
[Eu(hfa)$_3$(dpbp)]$_n$[d]	29	72	40	0.85	8.5×10^2	3.2×10^2

[a]$\lambda_{ex} = 380$ nm. [b]Eqs. (2.3)–(2.5). [c]$\lambda_{ex} = 355$ nm. [d]Ref. [24]

τ_{obs} and τ_{rad} are observed and radiative emission lifetimes, where τ_{rad} is defined as an ideal emission lifetime without non-radiative process. $A_{MD,0}$ is the spontaneous emission probability for $^5D_0 \rightarrow {}^7F_1$ transition in vacuo (14.65 s^{-1}), n is the refractive index of the medium (an average index of refraction equal to 1.5 is employed [31]), and I_{tot}/I_{MD} is the ratio of the total area of the corrected Eu(III) emission spectrum to the area of the $^5D_0 \rightarrow {}^7F_1$ band.

2.3 Results and Discussion

2.3.1 Photophysical Properties

The excitation, absorption, and emission spectra of the Eu(III) coordination polymers in solid state are shown in Fig. 2.3a. The coordination zippers exhibited bright red luminescence.

The photophysical parameters are summarized in Table 2.1. The intrinsic emission quantum yields (Φ_{ff}) for both polymers were estimated to be as high as that of the previous coordination polymer [Eu(hfa)$_3$(dpbp)]$_n$, indicating a low-vibrational coordination structure. The larger radiative rate constants (k_r) also reflect the greatly distorted coordination geometries around Eu(III) ions. The sharp signals at around 394, 416, 465, 472, 526, and 534 nm in excitation spectra were attributed to the 4f-4f transitions (Table 2.2). These spectroscopic features also supports the existence of allowed 4f-4f transitions in Eu(III) ions of these compounds, which relates to high intrinsic emission quantum yields. The most remarkable point is that both polymers exhibited twice as large Φ_{tot} values ($\Phi_{tot} \sim 60\%$), compared to [Eu(hfa)$_3$(dpbp)]$_n$ ($\Phi_{tot} = 29\%$).

The η_{sens} of [Eu(hfa)$_3$(dpt)]$_n$ and [Eu(hfa)$_3$(dpedot)]$_n$ were estimated to be 80 and 66%, respectively. The author considered the high η_{sens} to be responsible for the formation of low-lying ILCT states induced by the dense-packed zig-zag arrangement. In order to confirm the formation of ILCT states, an absorption spectrum in methanol and diffuse reflectance spectra in solid state were measured (Fig. 2.3b). In 10^{-5} M methanol solution, π-π* transition bands of hfa is observed at around 300 nm. In

2.3 Results and Discussion

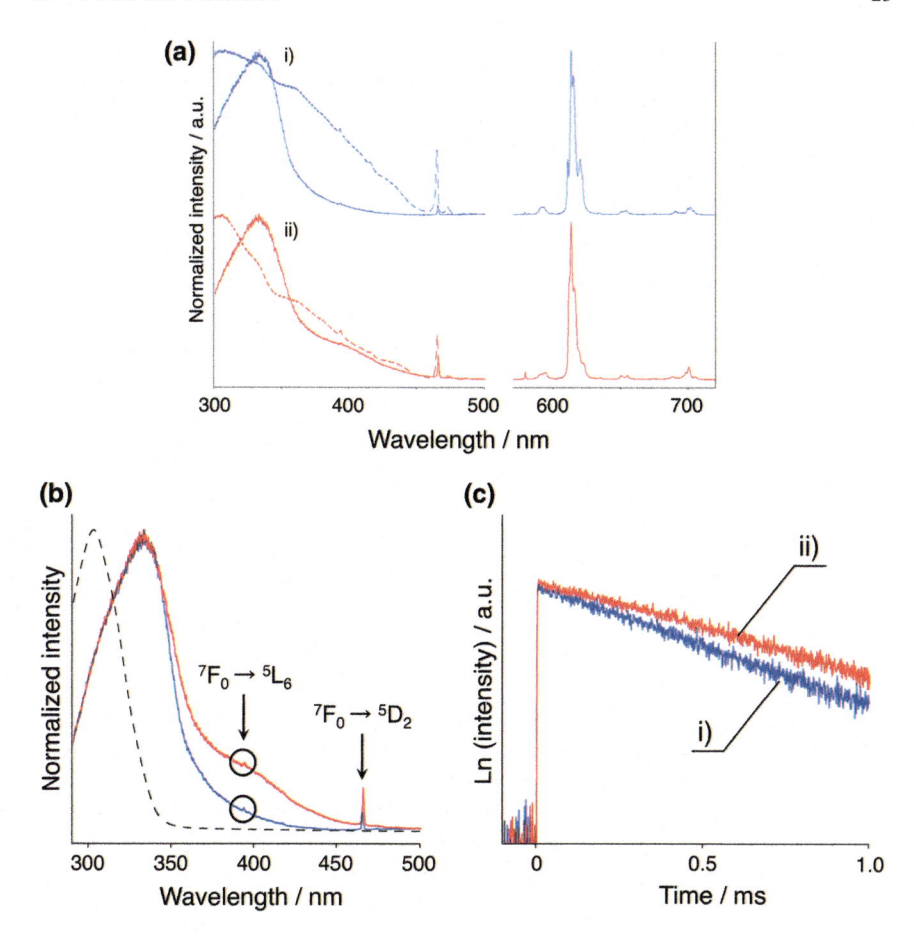

Fig. 2.3 **a** Diffuse reflectance (solid line), excitation (dotted line, monitored at 613 nm), and emission (solid line, $\lambda_{ex} = 380$ nm) spectra. **b** An absorption spectrum (10^{-5} M in MeOH for monomer, dashed line) and diffuse reflectance spectra (solid line). **c** Emission decay profiles ($\lambda_{ex} = 355$ nm) of (i) [Eu(hfa)$_3$(dpt)]$_n$ (blue line) and (ii) [Eu(hfa)$_3$(dpedot)]$_n$ (red line)

solid state, both polymers exhibited small 4f-4f absorption of Eu(III) ions at 394 nm and 465 nm. The broad absorption over UV to VIS regions was observed due to the ligand-based transitions, including π-π* absorption bands at 330 nm and ILCT bands at around 400 nm. The excitation spectra also indicate that the long-wavelength bands also contribute to Eu(III)-centered luminescence.

Table 2.2 The list of observed peaks in excitation spectra of [Eu(hfa)$_3$(X)]$_n$ (X = dpt, dpedot) and corresponding 4f-4f transitions that are generally found in absorption spectra of Eu(III) coordination compounds [32]

Transitions in excitation spectra	Transitions in absorption spectra		
Wavelength/nm	Transitions	Dipole character	Wavelength/nm
394	$^5L_6 \leftarrow {^7F_0}$	ED	390–405
416	$^5D_3 \leftarrow {^7F_1}$	MD	410–420
465	$^5D_2 \leftarrow {^7F_0}$	ED	460–470
472	$^5D_2 \leftarrow {^7F_1}$	ED	470–480
526	$^5D_1 \leftarrow {^7F_0}$	ED	520–530
534	$^5D_1 \leftarrow {^7F_1}$	ED	530–540

ED induced magnetic dipole transition, *MD* magnetic dipole transition

2.3.2 Coordination Structures

The η_{sens} should be related to the formation of ILCT states in solid state. Therefore, single crystals of both polymers were prepared to investigate the coordination structures. The crystal structures were determined to be typical 8-coordination with three hfa and two phosphine oxide ligands (Figs. 2.4a and 2.5a). First, the coordination geometry around Eu(III) ions was examined on the basis of crystal data. The shape factor S was calculated, and the observed dihedral angles (δ_i), ideal dihedral angles for square antiprism (θ_{SAP}) and trigonal dodecahedron (θ_{TDH}), calculated measure shape criteria, S_{SAP} and S_{TDH} are summarized (Fig. 2.6 and Table 2.3 for [Eu(hfa)$_3$(dpt)]$_n$, Fig. 2.7 and Table 2.4 for [Eu(hfa)$_3$(dpedot)]$_n$). When assuming 8-SAP structure for [Eu(hfa)$_3$(dpedot)]$_n$, the S value is much smaller than that for 8-TDH ($S_{SAP} = 5.13 < S_{TDH} = 10.6$), indicating that the coordination geometry is closer to 8-SAP than to 8-TDH. The geometry of [Eu(hfa)$_3$(dpt)]$_n$ can be defined as distorted 8-SAP structure due to the close S values ($S_{SAP} = 8.84$, $S_{TDH} = 9.21$).

The arrangement and intermolecular interactions of polymer chains were also investigated as shown in Figs. 2.4b and 2.5b. [Eu(hfa)$_3$(dpt)]$_n$ showed highly ordered and dense-packed structure due to the alternate arrangement of CF$_3$ and phenyl groups between single polymer chains. The multiple intermolecular CH/F [$d_{CH/F} = 2.80$ Å (H72/F9), 2.91 Å (H52/F10), 2.81 Å (H54/F6)] and CH/π [$d_{CH/\pi} = 2.79$ Å (H51/π), 2.91 Å (H53/π)] interactions were observed. In the case of [Eu(hfa)$_3$(dpedot)]$_n$, intramolecular π/π interactions were identified ($d_{\pi/\pi} = 3.56$ Å). The number of intermolecular CH/F interactions is much smaller than that for [Eu(hfa)$_3$(dpt)]$_n$, since the [Eu(hfa)$_3$(dpedot)]$_n$ forms relatively disordered polymer chains. The binding energies of CH/F and CH/π interactions are generally known to be 10–30 and 2–10 kJ mol^{-1}, respectively [33]. Thus, [Eu(hfa)$_3$(dpt)]$_n$ is highly zipped and shows large crystal density (Table 2.5).

2.3 Results and Discussion

(a)

(b)

CH/π interaction

H72 H53 H54
F9 H51 H52
F10 F6

CH/F interaction

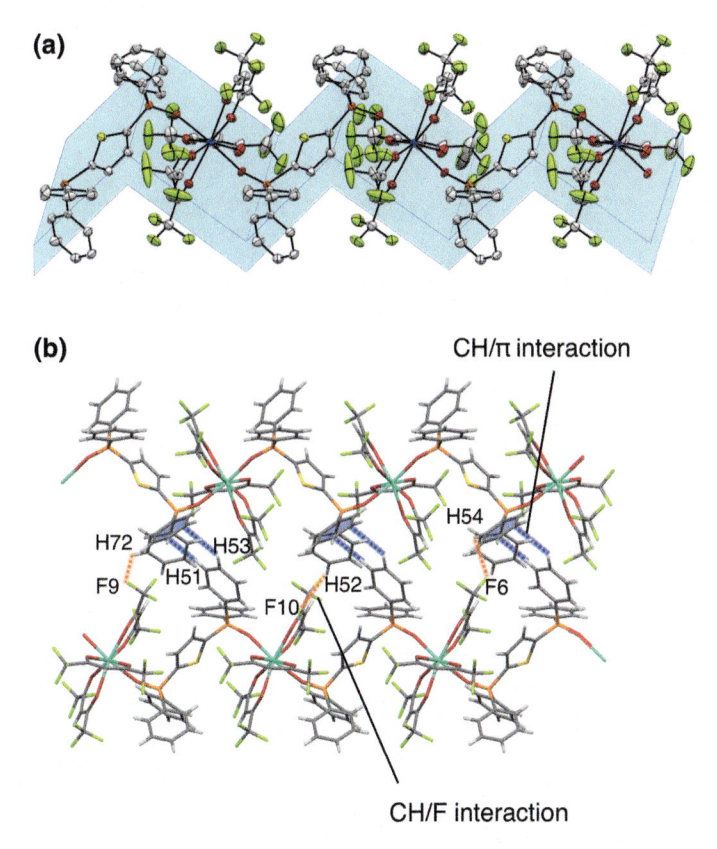

Fig. 2.4 a ORTEP drawing (showing 50% probability displacement ellipsoids) and **b** crystal packing structure focused on intermolecular interactions between single polymer chains of [Eu(hfa)$_3$(dpt)]$_n$

The distances between hfa ligands were estimated to evaluate the density of hfa ligands (Fig. 2.8). The intra-unit, intra-chain, and inter-chain hfa distances are shown in Table 2.6. The intra-chain distances are close in [Eu(hfa)$_3$(dpt)]$_n$ and [Eu(hfa)$_3$(dpedot)]$_n$ (smallest: 9 Å, largest: 14 Å). On the other hand, [Eu(hfa)$_3$(dpedot)]$_n$ and [Eu(hfa)$_3$(dpbp)]$_n$ exhibited close inter-chain distances (smallest: 7 Å, largest: 16 Å). These results indicate that intra-chain distances greatly affect η_{sens} values. Thus, introduction of thiophene-based bridging ligands can reduce the intra-chain hfa-hfa distances, which is ideal for efficient ligand-to-metal energy transfer. Bünzli and Eliseeva also reported ILCT-mediated high energy transfer efficiency (η_{sens} = 80%) of an Eu(III) dimer complex with close-packed hfa ligands [26].

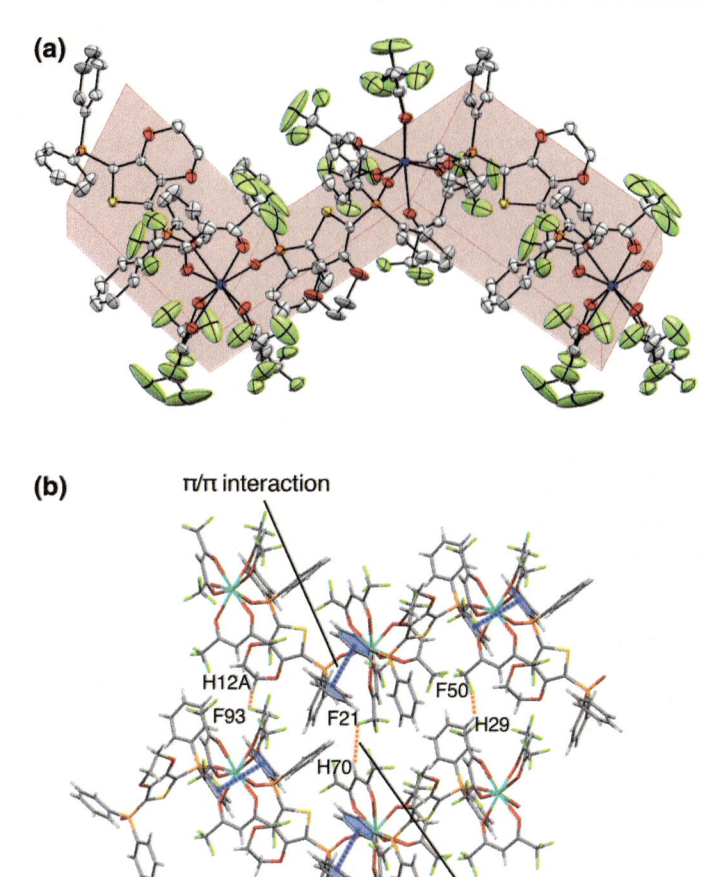

Fig. 2.5 a ORTEP drawing (showing 50% probability displacement ellipsoids) and **b** crystal packing structure focused on intermolecular interactions between single polymer chains of [Eu(hfa)$_3$(dpedot)]$_n$

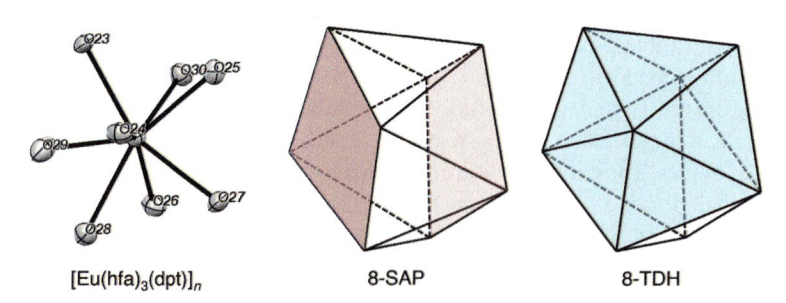

Fig. 2.6 Coordination environments around an Eu(III) ion of [Eu(hfa)$_3$(dpt)]$_n$

2.3 Results and Discussion

Table 2.3 The observed (δ_i), idealized dihedral angles ($\theta_{SAP}, \theta_{TDH}$), and calculated S values (S_{SAP}, S_{TDH}) for [Eu(hfa)$_3$(dpt)]$_n$

	δ_i	SAP				TDH		
		θ_{SAP}	$\delta_i - \theta_{SAP}$	$(\delta_i - \theta_{SAP})^2$		θ_{TDH}	$\delta_i - \theta_{TDH}$	$(\delta_i - \theta_{TDH})^2$
O24–O29	5.24	0	5.24	27.46		29.86	−24.62	606.14
O24–O23	68.88	77.1	−8.22	67.57		61.48	7.4	54.76
O24–O28	81.63	77.1	4.53	20.52		74.29	7.34	53.88
O29–O23	80.44	77.1	3.34	11.156		74.29	6.15	37.83
O29–O28	75.37	77.1	−1.73	2.99		61.48	13.89	192.93
O27–O30	19.75	0	19.75	390.06		29.86	−10.11	102.21
O27–O26	66.84	77.1	−10.26	105.27		61.48	5.36	28.73
O27–O25	68.77	77.1	−8.33	69.39		74.29	−5.52	30.47
O30–O25	72.46	77.1	−4.64	21.53		76.62	−4.16	17.31
O30–O26	75.09	77.1	−2.01	4.04		74.84	0.25	0.06
O26–O29	62.58	51.6	10.98	120.56		54.51	8.07	65.12
O26–O28	48.82	51.6	−2.78	7.73		48.09	0.73	0.53
O27–O24	38.83	51.6	−12.77	163.07		47.76	−8.93	79.74
O27–O28	54.06	51.6	2.46	6.05		54.48	−0.42	0.18
O25–O24	62.44	51.6	10.84	117.51		52.57	9.87	97.42
O25–O23	48.77	51.6	−2.83	8.01		52.85	−4.08	16.65
O30–O29	36.18	51.6	−15.42	237.78		47.34	−11.16	124.56
O30–O23	56.74	51.6	5.14	26.42		52.63	4.11	16.89
		$S_{SAP} = 8.84$				$S_{TDH} = 9.21$		

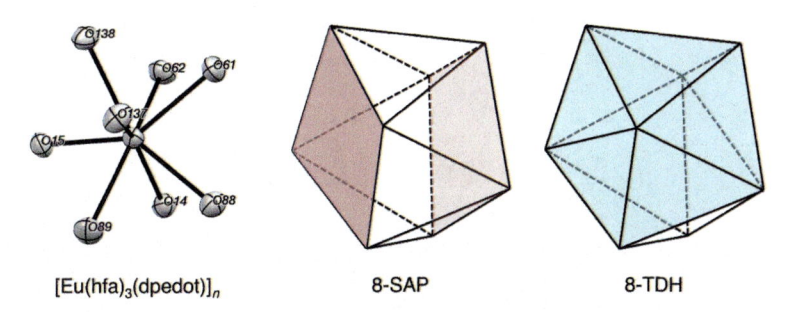

Fig. 2.7 Coordination environments around an Eu(III) ion of [Eu(hfa)$_3$(dpedot)]$_n$

2.3.3 Thermal Properties

Thermogravimetric analyses were carried out to determine the thermal stability of Eu(III) coordination zippers (Fig. 2.9). Thermal decomposition point of [Eu(hfa)$_3$(dpt)]$_n$ was estimated to be 322 °C, which was higher than [Eu(hfa)$_3$(dpbp)]$_n$ (308 °C) [24]. High thermal stability of [Eu(hfa)$_3$(dpt)]$_n$ was due to the dense-packed structures between single polymer chains with multiple intermolecular CH/F and CH/π interactions. [Eu(hfa)$_3$(dpedot)]$_n$ exhibited relatively low

Table 2.4 The observed (δ_i), idealized dihedral angles ($\theta_{SAP}, \theta_{TDH}$), and calculated S values (S_{SAP}, S_{TDH}) for [Eu(hfa)$_3$(dpedot)]$_n$

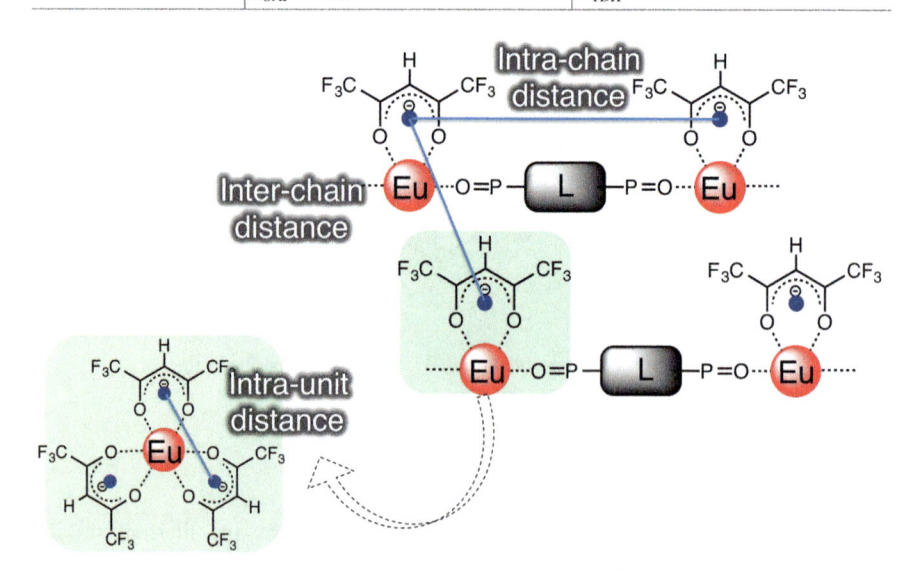

	δ_i	SAP				TDH		
		θ_{SAP}	$\delta_i - \theta_{SAP}$	$(\delta_i - \theta_{SAP})^2$		θ_{TDH}	$\delta_i - \theta_{TDH}$	$(\delta_i - \theta_{TDH})^2$
O137–O15	9.51	0	9.51	90.44		29.86	−20.35	414.12
O137–O138	73.01	77.1	−4.09	16.73		61.48	11.53	132.94
O137–O89	72.78	77.1	−4.32	18.66		74.29	−1.51	2.28
O15–O138	73.68	77.1	−3.42	11.70		74.29	−0.61	0.37
O15–O89	76.15	77.1	−0.95	0.90		61.48	14.67	215.21
O88–O62	2.24	0	2.24	5.02		29.86	−27.62	762.86
O88–O14	77.47	77.1	0.37	0.14		61.48	15.99	255.68
O88–O61	77.83	77.1	0.73	0.53		74.29	3.54	12.53
O62–O61	75.47	77.1	−1.63	2.66		76.62	−1.15	1.32
O62–O14	82.34	77.1	5.24	27.46		74.84	7.5	56.25
O14–O15	53.39	51.6	1.79	3.20		54.51	−1.12	1.25
O14–O89	47.91	51.6	−3.69	13.62		48.09	−0.18	0.03
O88–O137	44.55	51.6	−7.05	49.70		47.76	−3.21	10.30
O88–O89	60.99	51.6	9.39	88.17		54.48	6.51	42.38
O61–O137	54.42	51.6	2.82	7.95		52.57	1.85	3.42
O61–O138	46.58	51.6	−5.02	25.20		52.85	−6.27	39.31
O62–O15	44.37	51.6	−7.23	52.27		47.34	−2.97	8.82
O62–O138	59.28	51.6	7.68	58.98		52.63	6.65	44.22
		$S_{SAP} = 5.13$				$S_{TDH} = 10.6$		

Fig. 2.8 A schematic representation of the distances between hfa ligands in Eu(III) coordination polymers

2.3 Results and Discussion

Table 2.5 Crystallographic data of the Eu(III) coordination polymers

	[Eu(hfa)$_3$(dpt)]$_n$	[Eu(hfa)$_3$(dpedot)]$_n$	[Eu(hfa)$_3$(dpbp)]$_n$ [25]
Chemical formula	C$_{43}$H$_{25}$EuF$_{18}$O$_8$P$_2$S	C$_{45}$H$_{27}$EuF$_{18}$O$_{10}$P$_2$S	C$_{52}$H$_{31}$EuF$_{18}$O$_8$P$_2$
Formula weight	1257.60	1315.64	1339.69
Crystal system	Monoclinic	Triclinic	Monoclinic
Space group	$P2_1/n$ (#14)	$P1$ (#1)	$C2/c$ (#15)
a/Å	10.4933(5)	12.2292(14)	23.477(4)
b/Å	21.9927(11)	12.9103(11)	13.367(2)
c/Å	20.4330(11)	18.1033(19)	17.168(3)
α/deg	90.000	89.739(5)	90.000
β/deg	90.654(5)	84.208(4)	95.5749(8)
γ/deg	90.000	63.809(7)	90.000
Volume/Å^3	4715.1(4)	2549.1(5)	5340.5(16)
Z	4	2	4
d_{calc}/g cm^{-3}	1.771	1.714	1.666
Temperature/°C	−180	−150	−123
μ (Mo K$_\alpha$)/cm^{-1}	15.619	14.516	13.472
max 2θ/deg	55.0	55.0	55.0
Reflections collected	43336	25115	21091
Independent reflections	10696	18962	6099
R_1[a]	0.0492	0.0359	0.0266
wR_2[b]	0.1105	0.0954	0.0668

[a] $R_1 = \sum ||F_o| - |F_c||/\sum |F_o|$. [b] $wR_2 = [\sum w (F_o^2 - F_c^2)^2/\sum w (F_o^2)^2]^{1/2}$

Table 2.6 Distances between hfa ligands in Eu(III) coordination polymers

	[Eu(hfa)$_3$(dpt)]$_n$		[Eu(hfa)$_3$(dpedot)]$_n$		[Eu(hfa)$_3$(dpbp)]$_n$	
η_{sens}/%	80		66		40	
Intra-unit distance/Å	4.88		4.70		4.67	
	4.41		5.20		4.67	
	5.31		3.54		5.34	
	Average	4.87	Average	4.48	Average	4.89
Intra-chain distance/Å	Smallest	8.90	Smallest	9.19	Smallest	10.24
	Largest	14.1	Largest	14.4	Largest	18.0
Inter-chain distance/Å	Smallest	5.76	Smallest	7.23	Smallest	7.27
	Largest	9.98	Largest	15.7	Largest	15.8

Fig. 2.9 TGA thermograms of **a** [Eu(hfa)₃(dpt)]ₙ (blue line) and **b** [Eu(hfa)₃(dpedot)]ₙ (red line) under an argon atmosphere (5 °C min⁻¹)

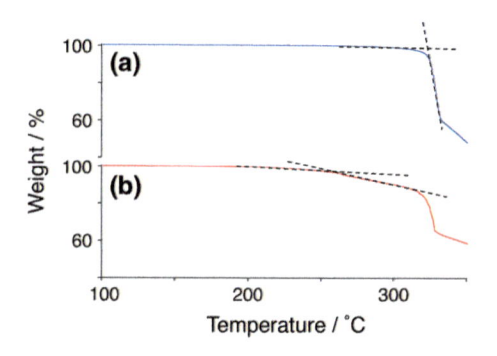

decomposition temperature (264 °C). A small drop in the thermogravimetric curve is assumed to be responsible for the degradation of dioxane ring in dpedot ligands, followed by desorption and degradation of hfa ligands.

2.3.4 DFT Calculations

The dipole moment D of the bridging ligands was also estimated using DFT calculations [B3LYP 6-31G**] on the basis of CIF data (Fig. 2.10). Compared to the previously reported compounds, the bridging ligands in coordination zippers have large D values.

The polar character of bridging ligands probably results in the characteristic alternate arrangement of inter- or intra-polymer chains. Eu(III) coordination polymers with larger D values were also found to show larger Φ_{ff}. Mason and co-workers proposed the "ligand polarization" theory of 4f-4f hypersensitivity, that described the relationship between induced electric dipole moment and dynamic coupling in the Judd-Offelt theory [34, 35]. The effect on dipole moment of the ligand in an Ln(III) complex also has been reported [36, 37]. The 4f-4f transition probability should be increased by the ligands with large dipole moments, since the polarizability generally describes the strength of field-induced dipoles, and the compounds with large ground state dipole moments tend to exhibit large induced dipole moments. The enhancement of Φ_{ff} in Eu(III) coordination zippers were thus caused by the larger magnitude of the dipole moment in bridging ligands.

2.4 Conclusions

In conclusion, highly emissive and thermally stable Eu(III) coordination polymers [Eu(hfa)₃(dpt)]ₙ and [Eu(hfa)₃(dpedot)]ₙ were successfully synthesized by introducing thiophene-based bridges. They exhibit bright red luminescence with energy

2.4 Conclusions

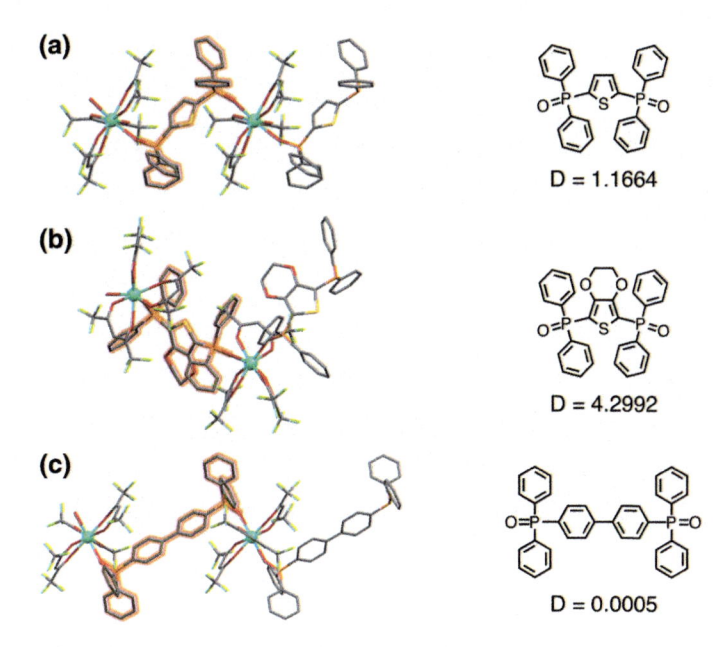

Fig. 2.10 Crystal structures of Eu(III) coordination polymers focused on bridging ligands for DFT calculations (left) and chemical structures and calculated D values (right) for **a** dpt, **b** dpedot, and **c** dpbp

transfer efficiency of up to 80%. Efficient ligand-to-metal energy transfer was due to the formation of dense-packed coordination zipper structures induced by the small and bent bridging ligands. Incorporation of polar thiophene-based bridges was also found to enhance the 4f-4f transition of Eu(III) ions.

Along with the conventional molecular design of ligand fields around Eu(III) ions, novel guidelines for a dense-packed assembly of luminescent Ln(III) coordination polymers were provided. The reported strategy for coordination zippers would be advantageous for efficient energy transfer and strong luminescence in the solid state, which can be employed in highly emissive and stable materials for optical applications.

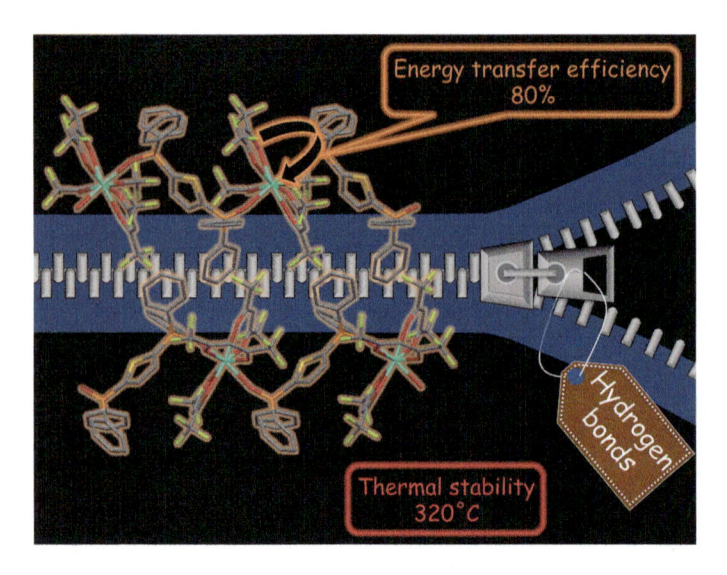

References

1. J.H. Burroughes, D.D.C. Bradley, A.R. Brown, R.N. Marks, K. Mackay, R.H. Friend, P.L. Burn, A.B. Holmes, Nature **347**, 539–541 (1990)
2. C.D. Dimitrakopoulos, P.R.L. Malenfant, Adv. Mater. **14**, 99–117 (2002)
3. E.G. Moore, A.P.S. Samuel, K.N. Raymond, Acc. Chem. Res. **42**, 542–552 (2009)
4. S.V. Eliseeva, J.-C.G. Bünzli, Chem. Soc. Rev. **39**, 189–227 (2010)
5. M. Schaferling, Angew. Chem. Int. Ed. **51**, 3532–3554 (2012)
6. J.F. Callan, A.P. de Silva, D.C. Magri, Tetrahedron **61**, 8551–8588 (2005)
7. A.P. de Silva, H.Q.N. Gunaratne, T. Gunnlaugsson, A.J.M. Huxley, C.P. McCoy, J.T. Rademacher, T.E. Rice, Chem. Rev. **97**, 1515–1566 (1997)
8. S.W. Thomas, G.D. Joly, T.M. Swager, Chem. Rev. **107**, 1339–1386 (2007)
9. C. Adachi, M.A. Baldo, M.E. Thompson, S.R. Forrest, J. Appl. Phys. **90**, 5048–5051 (2001)
10. A. de Bettencourt-Dias, Dalton Trans. **22**, 2229–2241 (2007)
11. J.-C.G. Bünzli, C. Piguet, Chem. Soc. Rev. **34**, 1048–1077 (2005)
12. K. Binnemans, Chem. Rev. **109**, 4283–4374 (2009)
13. T. Gunnlaugsson, M. Glynn, G.M. Tocci, P.E. Kruger, F.M. Pfeffer, Coord. Chem. Rev. **250**, 3094–3117 (2006)
14. G.E. Khalil, K. Lau, G.D. Phelan, B. Carlson, M. Gouterman, J.B. Callis, L.R. Dalton, Rev. Sci. Instrum. **75**, 192–206 (2004)
15. N.B.D. Lima, S.M.C. Goncalves, S.A. Junior, A.M. Simas, Sci. Rep. **3** (2013)
16. A. de Bettencourt-Dias, P.S. Barber, S. Viswanathan, Coord. Chem. Rev. **273**, 165–200 (2014)
17. L. Armelao, S. Quici, F. Barigelletti, G. Accorsi, G. Bottaro, M. Cavazzini, E. Tondello, Coord. Chem. Rev. **254**, 487–505 (2010)
18. K. Binnemans, R. Van Deun, C. Gorller-Walrand, S.R. Collinson, F. Martin, D.W. Bruce, C. Wickleder, Phys. Chem. Chem. Phys. **2**, 3753–3757 (2000)
19. M.H.V. Werts, R.T.F. Jukes, J.W. Verhoeven, Phys. Chem. Chem. Phys. **4**, 1542–1548 (2002)
20. H.B. Zhang, L.J. Zhou, J. Wei, Z.H. Li, P. Lin, S.W. Du, J. Mater. Chem. **22**, 21210–21217 (2012)

References

21. M.S. Liu, Q.Y. Yu, Y.P. Cai, C.Y. Su, X.M. Lin, X.X. Zhou, J.W. Cai, Cryst. Growth Des. **8**, 4083–4091 (2008)
22. J. Rocha, L.D. Carlos, F.A.A. Paz, D. Ananias, Chem. Soc. Rev. **40**, 926–940 (2011)
23. S.V. Eliseeva, D.N. Pleshkov, K.A. Lyssenko, L.S. Lepnev, J.-C.G. Bünzli, N.P. Kuzmina, Inorg. Chem. **49**, 9300–9311 (2010)
24. K. Miyata, T. Ohba, A. Kobayashi, M. Kato, T. Nakanishi, K. Fushimi, Y. Hasegawa, ChemPlusChem **77**, 277–280 (2012)
25. A. D'Aleo, F. Pointillart, L. Ouahab, C. Andraud, O. Maury, Coord. Chem. Rev. **256**, 1604–1620 (2012)
26. S.V. Eliseeva, O.V. Kotova, F. Gumy, S.N. Semenov, V.G. Kessler, L.S. Lepnev, J.-C.G. Bünzli, N.P. Kuzmina, J. Phys. Chem. A **112**, 3614–3626 (2008)
27. E.R. Trivedi, S.V. Eliseeva, J. Jankolovits, M.M. Olmstead, S. Petoud, V.L. Pecoraro, J. Am. Chem. Soc. **136**, 1526–1534 (2014)
28. Y. Hasegawa, R. Hieda, K. Miyata, T. Nakagawa, T. Kawai, Eur. J. Inorg. Chem. **32**, 4978–4984 (2011)
29. J.D. Xu, E. Radkov, M. Ziegler, K.N. Raymond, Inorg. Chem. **39**, 4156–4164 (2000)
30. A. Aebischer, F. Gumy, J.-C.G. Bünzli, Phys. Chem. Chem. Phys. **11**, 1346–1353 (2009)
31. R. Pavithran, N.S.S. Kumar, S. Biju, M.L.P. Reddy, S.A. Junior, R.O. Freire, Inorg. Chem. **45**, 2184–2192 (2006)
32. K. Binnemans, Coord. Chem. Rev. **295**, 1–45 (2015)
33. G.R. Desiraju, T. Steiner, *The Weak Hydrogen Bond in Structural Chemistry and Biology* (Oxford Univ. Press, 1999)
34. S.F. Mason, R.D. Peacock, B. Stewart, Chem. Phys. Lett. **29**, 149–153 (1974)
35. S.F. Mason, R.D. Peacock, B. Stewart, Mol. Phys. **30**, 1829–1841 (1975)
36. T. Nakagawa, Y. Hasegawa, T. Kawai, J. Phys. Chem. A **112**, 5096–5103 (2008)
37. Y. Hasegawa, N. Sato, Y. Hirai, T. Nakanishi, Y. Kitagawa, A. Kobayashi, M. Kato, T. Seki, H. Ito, K. Fushimi, J. Phys. Chem. A **119**, 4825–4833 (2015)

Chapter 3
Luminescent Lanthanide-Mixed Coordination Polymers for Tunable Temperature-Sensitivity

Abstract The control of energy transfer efficiency in lanthanide [Ln(III)]-mixed coordination polymers is reported. The coordination polymers [Tb,Eu(hfa)$_3$(dpbp)]$_n$ are composed of Tb(III) ions, Eu(III) ions, hfa ligands, and bidentate phosphine oxide ligands [dpbp: 4,4′-bis(diphenylphosphoryl)biphenyl]. The emission colors were controlled by varying the mixture ratio of Tb(III) and Eu(III) ions (Tb/Eu = 1–1000). The obtained compounds were characterized by XRD, emission spectra, and emission lifetime measurements. Temperature-dependent emission color change from green, yellow, orange, to red was observed, and spectroscopic features were discussed on the basis of energy transfer efficiency in the solid state.

Keywords Lanthanide · Coordination polymer · Temperature sensitivity
Thermal stability

3.1 Introduction

In the field of fluid dynamics, visualization of flow fields is a key technique for the development of efficient transportation systems such as cars, ships, and aircrafts [1–5]. Flow visualization can provide a straightforward and qualitative assessment of fluid flow on the surface of a material. Some visualization methods using temperature- or pressure-sensitive paints (TSP or PSP, Fig. 3.1) are called molecular imaging techniques and they provide detailed information [6–9]. These techniques have the advantage of describing the entire flow field on objects [10, 11] unlike discrete point probes such as thermocouples and piezoelectric pressure sensors that provide point-by-point information with insufficient spatial resolution. For unsteady flows in particular, fluid visualization is known to render details of the flow field far more quickly than conventional measurements [12].

In molecular imaging techniques, temperature and pressure on materials' surface have been detected by thermal and oxygen quenching of the luminescence from organic compounds and metal complexes [8, 9]. However, these compounds intrinsically show broad, mono-colored, weak emission and relatively low decomposition

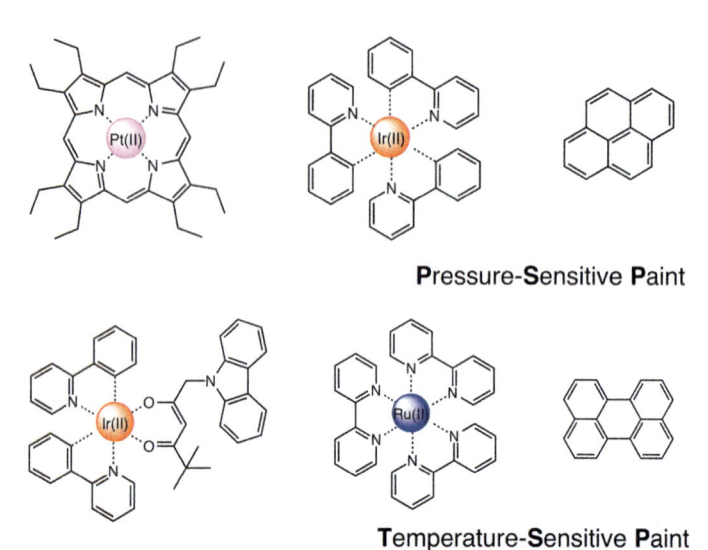

Fig. 3.1 Chemical structures of luminescent compounds used in pressure sensitive paint (PSP) and temperature sensitive paint (TSP)

points (<200 °C), resulting in a limited temperature-sensitive region. Thermally-stable and temperature-sensitive dyes are highly desired for temperature/pressure detection of atmospheric vehicles and hypersonic jetliners that are subjected to atmospheric re-entry which accompanies a significant temperature elevation due to aerodynamic heating in the hypersonic flow.

Luminescent physical sensors based on Ln(III) coordination compounds are attractive in terms of the resolution of signal detection due to their characteristic sharp and long-lived emission [13–22]. In 2003, Amao and co-workers first developed luminescent Eu(III) complexes with temperature sensitivity [23]. Wolfbeis demonstrated that Eu(III) β-diketonate complexes can be used for high-resolution oxygen profiles [24]. Katagiri reported a Tb(III) complex with effective energy back transfer (BET) from the emitting level of Tb(III) ions to the excited triplet state of hfa ligands, which is suitable for temperature sensors [25]. Hasegawa et al. also reported a Tb(III)/Eu(III) mixed system with hfa ligands, $[Tb,Eu(hfa)_3(dpbp)]_n$ (Tb/Eu = 99), that exhibits dramatic color change depending on the temperature from 200 to 500 K (Fig. 3.2a) [26]. This is due to the BET as well as energy transfer between Ln(III) ions (Fig. 3.2b). Thermal stability was also introduced by intermolecular CH/π and CH/F hydrogen bonds. Since Tb(III) ions are energy donors and Eu(III) ions are energy acceptors in these "chameleon luminophore" systems, energy transfer efficiency between lanthanide ions can be tuned by changing the Tb(III)/Eu(III) mixture ratios. The varied energy transfer efficiency should lead to variable emission colors and temperature-sensitive regions.

3.1 Introduction

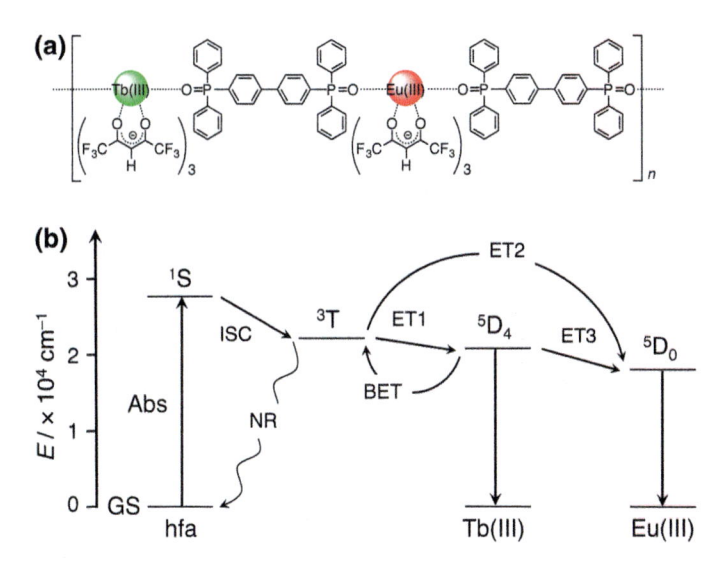

Fig. 3.2 **a** A chemical structure of Tb(III)/Eu(III) mixed coordination polymer and **b** corresponding energy transfer diagram [26]

Photophysical studies on the energy transfer from Tb(III) to Eu(III) ions are common in mono- and di-nuclear lanthanide complexes in homogeneous organic and aqueous media [27, 28]. Recently, carboxylate-based Ln(III)-mixed coordination polymers including MOFs have been extensively developed for luminescent thermometers over a wide temperature range [29–32]. Temperature sensitive region generally depends on organic bridging ligands, and study on the effect of Ln(III) mixture ratios on temperature sensitivity is still in progress. Kinetic analyses of the photophysical properties in solid-state coordination polymers are directly linked to the development of effective molecular thermometers for materials' surface.

In this chapter, temperature-dependent energy transfer efficiencies between Tb(III) and Eu(III) ions in coordination polymers are described to estimate the luminescence performance. The Tb(III)/Eu(III) mixed coordination polymers, [Tb,Eu(hfa)$_3$(dpbp)]$_n$, were prepared and characterized by IR and XRD measurements. The energy transfer efficiencies in the range between 100 and 400 K were evaluated on the basis of emission spectra and lifetime measurements in a cryostat. The characteristic photophysical properties were found in a coordination polymer with Tb(III) ions and a small amount of Eu(III) ions (Tb/Eu = 750).

3.2 Experimental Section

3.2.1 Materials

Europium(III) acetate *n*-hydrate (99.9%) and terbium(III) acetate tetrahydrate (99.9%) were purchased from Wako Pure Chemical Industries Ltd. 4,4′-Dibromobiphenyl, hexafluoroacetylacetone, chlorodiphenylphosphine and ytterbium(III) acetate tetrahydrate were obtained from Tokyo Chemical Industry Co. Ltd. and Aldrich Chemical Company Inc. All other chemicals and solvents were reagent grade and were used as received.

3.2.2 Apparatus

Infrared spectra were recorded on a JASCO FT/IR-420 spectrometer. Powder X-ray diffraction patterns were collected on a RIGAKU RINT 2000 Ultima. Obtained patterns were calibrated by the signal of silicon powder at 28.4°. ^{1}H NMR (400 MHz) spectrum was recorded on a JEOL ECS400. Chemical shifts are reported in δ ppm, referenced to an internal tetramethylsilane standard. Mass spectrum was measured using a JEOL JMS-700TZ. Elemental analysis was performed using a Yanaco CHN corderMT-6. Inductively coupled plasma (ICP) emission spectroscopy was performed on a Shimadzu ICPE-9000.

3.2.3 Syntheses

The organic bridging ligands and corresponding coordination polymers were synthesized following Scheme 3.1.

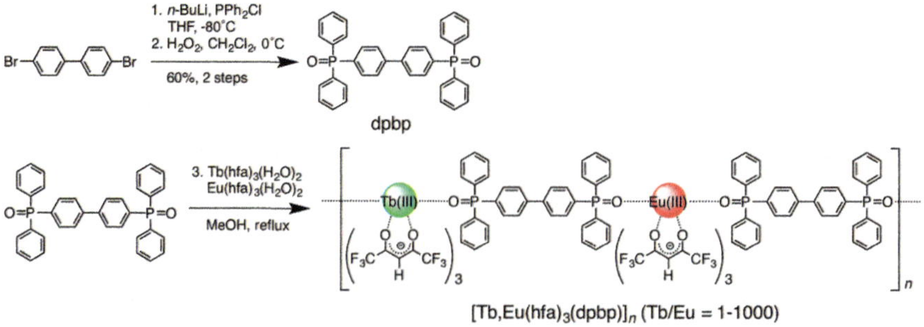

Scheme 3.1 Synthetic schemes of bridging ligand and Tb(III)/Eu(III) mixed coordination polymers

3.2 Experimental Section

Preparation of 4,4′-bis(diphenylphosphoryl)biphenyl (dpbp) [22]

4,4′-Bis(diphenylphosphoryl)biphenyl was synthesized according to the published procedure. 4,4′-Dibromobiphenyl (11 g, 36 mmol) was dissolved in dry THF (140 mL). A solution of 1.6 M n-BuLi (56 mL, 90 mmol) was added dropwise to the solution at −80 °C. The mixture was allowed to stir for 3 h at −10 °C, after which a PPh$_2$Cl (17 mL, 90 mmol) was added dropwise at −80 °C. The mixture was gradually brought to room temperature, and stirred overnight. The product was extracted with ethyl acetate, concentrated, and resulting residue was washed with acetone and ethanol for several times. The obtained white solid was dissolved in dichloromethane (120 mL), followed by addition of a 30% H$_2$O$_2$ aqueous solution (30 mL). The reaction mixture was stirred for 3 h at 0 °C. The product was extracted with dichloromethane, washed with brine for three times and dried over anhydrous MgSO$_4$. The solvent was evaporated to afford a white powder of 4,4′-bis(diphenylphsphoryl)biphenyl (dpbp).

Yield: 12 g (60%). ^{1}H NMR (400 MHz, CDCl$_3$, 25 °C) δ 7.66–7.77 (m, 16H, –CH), δ 7.53–7.57 (m, 4H, –CH), δ 7.44–7.49 (m, 8H, -CH) ppm. ESI-Mass (m/z): calcd for C$_{36}$H$_{29}$O$_2$P$_2$ [M+H]$^+$, 555.2; found, 555.2. Anal. Calcd for C$_{36}$H$_{28}$O$_2$P$_2$: C, 77.97; H, 5.09%. Found: C, 77.49; H, 5.20%.

Preparation of Tb(III)/Eu(III) mixed coordination polymers [Tb,Eu(hfa)$_3$(dpbp)]$_n$ (Tb/Eu = X).

Tb(hfa)$_3$(H$_2$O)$_2$ (x mg) and Eu(hfa)$_3$(H$_2$O)$_2$ (y mg) were dissolved in 30 mL of methanol (see Table 3.1 for x and y). The Tb(III)/Eu(III)-mixed solution (in total 0.80 mmol, 30 mL) was added to a methanol solution (30 mL) of dpbp ligand (0.44 g, 0.80 mL), and refluxed for 2 h to form white precipitates. The precipitates were filtered, washed with methanol and chloroform for several times, and dried in vacuo. The obtained compounds were identified using IR spectroscopy (Fig. 3.3).

IR (KBr): 1650 (st, C=O), 1253-1142 (st, C–F), 1125 (st, P=O) cm^{-1}.

Table 3.1 Tb(III)/Eu(III) mixture ratio for syntheses of [Tb,Eu(hfa)$_3$(dpbp)]$_n$ (Tb/Eu = X)

X	Tb(hfa)$_3$(H$_2$O)$_2$/x mg	Eu(hfa)$_3$(H$_2$O)$_2$/y mg
1	326	324
10	594	58.9
50	640	12.7
150	649	4.29
250	650	2.58
350	651	1.84
500	652	1.29
650	652	0.99
750	652	0.86
850	652	0.76
1000	652	0.65

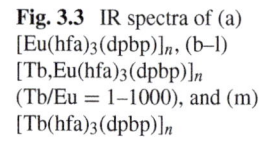

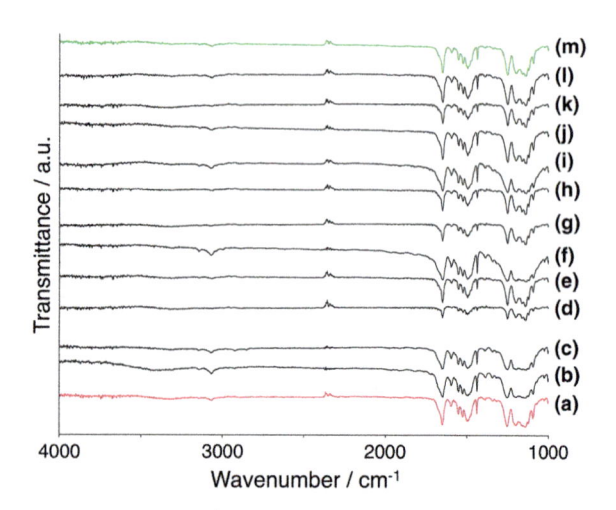

Fig. 3.3 IR spectra of (a) [Eu(hfa)₃(dpbp)]ₙ, (b–l) [Tb,Eu(hfa)₃(dpbp)]ₙ (Tb/Eu = 1–1000), and (m) [Tb(hfa)₃(dpbp)]ₙ

Table 3.2 Emission intensities and estimated Tb/Eu mixture ratios based on ICP measurements

(Tb/Eu)$_{syn}$[a]	I_{Tb}	I_{Eu}	(Tb/Eu)$_{ICP}$[b]
1	9769	9856.7	0.99
10	9568	973.41	9.83
50	9687	194.68	49.76
150	9854	64.89	151.86
250	9867	38.94	253.39
350	9876	27.81	355.12
500	9513	19.47	488.60
650	9687	14.98	646.66
750	9986	12.98	769.34
850	9956	11.45	869.52
1000	9947	9.73	1022.30

[a]Synthetic Tb/Eu mixture ratio. [b]Tb/Eu mixture ratio estimated from ICP analyses

ICP-AES was also conducted to determine Tb/Eu mixture ratios of [Tb,Eu(hfa)₃(dpbp)]ₙ. Thoroughly dried [Tb,Eu(hfa)₃(dpbp)]ₙ powders were digested in 60% HNO₃ aqueous solution, and then diluted to 1.2% HNO₃. Calibration standards were prepared by diluting 1000 ppm standards of both Eu and Tb (purchased from Kanto Chemical) over the range of 0.1–1000 ppb. Concentrations were calculated from the emission intensities of Eu and Tb (λ_{obs} = 381.917 and 350.917 nm, respectively) for each sample using the appropriate calibration curves and averaged. These analyses revealed Tb/Eu mixture ratios (Table 3.2).

3.2 Experimental Section

3.2.4 Optical Measurements

Emission spectra were recorded on a JASCO F-6300-H spectrometer and corrected for the response of the detector system. Emission lifetimes of Tb(III) ions in coordination polymers were measured using the third harmonics (355 nm) of a Q-switched Nd:YAG laser (Spectra Physics, INDI-50, fwhm = 5 ns, $\lambda = 1064$ nm), filter transmitting wavelength shorter than 550 nm, and photomultiplier (Hamamatsu photonics R5108, response time ≤ 1.1 ns). The Nd:YAG laser response was monitored with a digital oscilloscope (Sony Tektronix, TDS3052, 500 MHz) synchronized to the single-pulse excitation. Emission lifetimes were determined from the slope of logarithmic plots of the decay profiles. Emission lifetimes and emission spectra from the range between 100 and 400 K were measured with a cryostat (Thermal Block Company, SA-SB245T) and a temperature controller (Oxford, Instruments, ITC 502S). The emission quantum yields of coordination polymers excited at 380 nm were estimated using JASCO F-6300-H spectrometer attached with JASCO ILF-533 integrating sphere unit ($\varphi = 100$ mm). The wavelength dependences of the detector response and the beam intensity of Xe light source for each spectrum were calibrated using a standard light source.

3.3 Results and Discussion

3.3.1 Powder X-Ray Diffraction Measurements

The crystal structures of prepared Ln(III) coordination polymers were identified using powder XRD analyses. All signals were calibrated by the signal of silicon powder at 28.4°. Observed signals at 7.5°, 8.8°, 9.4°, 10.3°, 20.1° and 21.5° were attributed to the geometrical structures of [Tb,Eu(hfa)$_3$(dpbp)]$_n$ (Fig. 3.4). The signals of [Tb,Eu(hfa)$_3$(dpbp)]$_n$ were found to agree with those of previously reported [Eu(hfa)$_3$(dpbp)]$_n$ and [Tb(hfa)$_3$(dpbp)]$_n$ [22], indicating that the Tb(III)/Eu(III) mixture ratios had no effect on contraction or extension of the crystal lattice. Thus, the geometrical structures of [Tb,Eu(hfa)$_3$(dpbp)]$_n$ are the same as that of [Eu(hfa)$_3$(dpbp)]$_n$. According to the single crystal structure of [Eu(hfa)$_3$(dpbp)]$_n$, distances between Eu(III) ions of inter- and intra-polymer chains were 11.4 and 13.6 Å, respectively.

3.3.2 Emission Spectra

The photophysical properties of selected samples (Tb/Eu = 1, 10, 250, 500, 750) are discussed below. In order to estimate the energy transfer efficiency between Tb(III) and Eu(III) ions in solid state, temperature-dependent photophysical properties of [Tb,Eu(hfa)$_3$(dpbp)]$_n$ were investigated. [Tb,Eu(hfa)$_3$(dpbp)]$_n$ (Tb/Eu = 10) showed

Fig. 3.4 XRD patterns of (a) [Eu(hfa)$_3$(dpbp)]$_n$, (b–l) [Tb,Eu(hfa)$_3$(dpbp)]$_n$ (Tb/Eu = 1–1000), and (m) [Tb(hfa)$_3$(dpbp)]$_n$

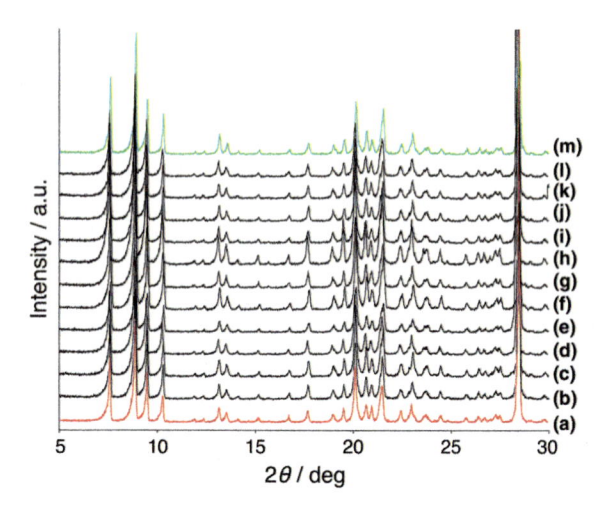

Fig. 3.5 Temperature-dependent emission spectra of [Tb,Eu(hfa)$_3$(dpbp)]$_n$ (λ_{ex} = 380 nm, Tb/Eu = 10)

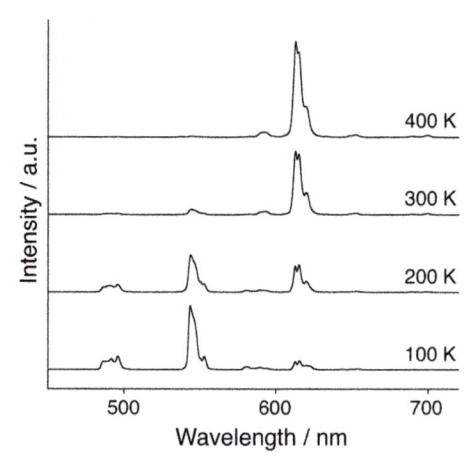

emission color change from green, yellow, orange, to red in the temperature range between 100 and 400 K, and obtained spectra were presented in Fig. 3.5.

The characteristic emission bands at 488, 543, and 613 nm are attributed to 4f-4f transitions of Tb(III) ($^5D_4 \rightarrow {}^7F_6$, $^5D_4 \rightarrow {}^7F_5$) and Eu(III) ($^5D_0 \rightarrow {}^7F_2$), respectively. The emission quantum yield and lifetime at 300 K (λ_{ex} = 380 nm) were 59.8% and 0.11 ms, respectively. The emission intensity of Tb(III) ions in [Tb,Eu(hfa)$_3$(dpbp)]$_n$ decreased with increasing temperature, particularly at temperatures above 300 K. On the contrary, the emission intensity of Eu(III) ions in [Tb,Eu(hfa)$_3$(dpbp)]$_n$ increased with rise of temperature. The emission intensity ratio of Tb(III) and Eu(III) ions (I_{Eu}/I_{Tb}) therefore exhibited a dramatic change, and the values for [Tb,Eu(hfa)$_3$(dpbp)]$_n$ (Tb/Eu = 1, 10, 250, 500, 750) were summarized in Fig. 3.6a. The intensity ratio (I_{Eu}/I_{Tb}) depended on the concentrations of Tb(III) and Eu(III) ions.

3.3 Results and Discussion

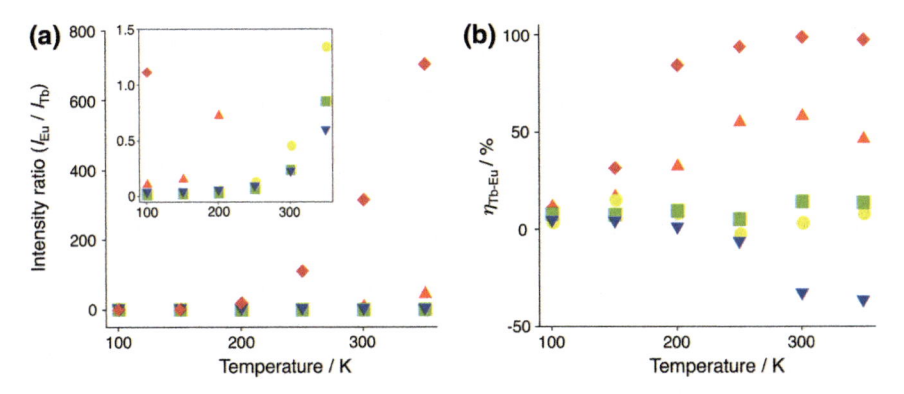

Fig. 3.6 Temperature dependence of **a** emission intensity ratios (I_{Eu}/I_{Tb}) and **b** energy transfer efficiencies of [Tb,Eu(hfa)$_3$(dpbp)]$_n$ (λ_{ex} = 380 nm, filled diamond: Tb/Eu = 1, filled triangle: Tb/Eu = 10, filled circle: Tb/Eu = 250, filled square: Tb/Eu = 500 and filled invert triangle: Tb/Eu = 750)

3.3.3 Energy Transfer Efficiency

The temperature-dependent emission spectra of [Tb,Eu(hfa)$_3$(dpbp)]$_n$ are due to the energy transfer between Tb(III) and Eu(III) ions. The energy transfer efficiency from Tb(III) to Eu(III) ions (η_{Tb-Eu}) are estimated by following Eq. (3.1) [33] when assuming that the energy transfer process is dipolar-dipolar,

$$\eta_{Tb-Eu} = 1 - \left(\frac{\tau_{obs}}{\tau_{Tb}}\right) \quad (3.1)$$

where τ_{obs} is the Tb ($^5D_4$) emission lifetime in a Tb(III)/Eu(III) mixed system ([Tb,Eu(hfa)$_3$(dpbp)]$_n$) and τ_{Tb} is the lifetime in a pure Tb(III) system ([Tb(hfa)$_3$(dpbp)]$_n$). The η_{Tb-Eu} value of [Tb,Eu(hfa)$_3$(dpbp)]$_n$ (Tb/Eu = 99) was reported to be 38% at room temperature [26]. The author here considers that the Eq. (3.1) is similarly applicable to the compounds in this study. The emission lifetimes were estimated by monitoring the 4f-4f transition bands of Tb(III) ions, $^5D_4 \rightarrow {}^7F_6$ (545 nm) and $^5D_4 \rightarrow {}^7F_5$ (488 nm), using an optical filter (550 nm short pass filter). The obtained emission decay profiles were analyzed as single exponential and summarized in Table 3.3.

The temperature-dependent η_{Tb-Eu} of [Tb,Eu(hfa)$_3$(dpbp)]$_n$ were shown in Fig. 3.6b. The η_{Tb-Eu} values at 100 K were approximately zero. According to the measurements at 300 K, the η_{Tb-Eu} were found to be 99, 65, 7.4, and 8.2% for Tb/Eu = 1, 10, 250, and 500, respectively. These results indicate that η_{Tb-Eu} depends on the concentration of the energy transfer acceptors, Eu(III) ions. [Tb,Eu(hfa)$_3$(dpbp)]$_n$ (Tb/Eu = 750) was also found to show negative η_{Tb-Eu} value (−33%) due to the longer emission lifetime of Tb/Eu = 750 mixture than that of Tb(III) pure one. The author considers this to be responsible for the different dominating concentration quenching mechanisms of Tb(III) and Eu(III) ions.

Table 3.3 Emission lifetimes of Tb(III) ions in chameleon polymers

Temperature /K	[Tb,Eu(hfa)$_3$(dpbp)]$_n$ (Tb/Eu = X)					[Tb(hfa)$_3$(dpbp)]$_n$
	X = 1	X = 10	X = 250	X = 500	X = 750	
100	0.85	0.87	0.89	0.90	0.92	0.95
150	0.65	0.82	0.87	0.90	0.93	0.94
200	0.11	0.58	0.79	0.79	0.85	0.83
250	0.024	0.26	0.62	0.60	0.69	0.63
300	0.0034	0.11	0.30	0.28	0.40	0.30
350	0.0027	0.054	0.11	0.12	0.17	0.12

3.4 Conclusions

Lanthanide coordination polymers with various ionic ratios, [Tb,Eu(hfa)$_3$(dpbp)]$_n$ (Tb/Eu = 1, 10, 250, 500, 750), were synthesized and temperature-dependent photophysical properties were examined. The energy transfer efficiency between Ln(III) ions was depended on Tb(III)/Eu(III) mixture ratios, and emission color tuning was also demonstrated. Characteristic luminescent properties of lanthanide coordination polymers composed of Tb(III) and a small amount of Eu(III) ions (Tb/Eu = 750) were also observed. In the solid-state polymers, the bridging ligands also affect the thermo-sensing properties. These studies will provide information on novel aspects of the energy transfer mechanisms in solid-state lanthanide coordination polymers and lead to further development of ideal sensor materials

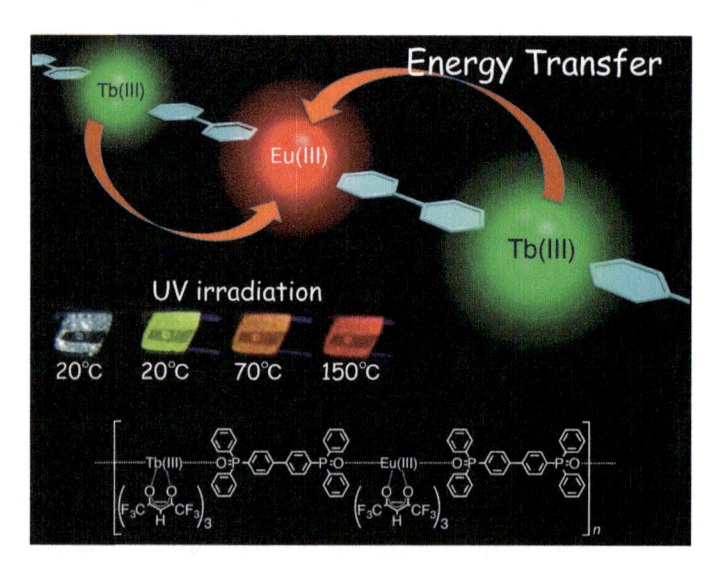

References

1. R.J. Adrian, Annu. Rev. Fluid Mech. **23**, 261–304 (1991)
2. J.H. Bell, E.T. Schairer, L.A. Hand, R.D. Mehta, Annu. Rev. Fluid Mech. **33**, 155–206 (2001)
3. W.L. Barth, C.A. Burns, IEEE Trans. Visual Comput. Graphics **13**, 1751–1758 (2007)
4. H. Sakaue, T. Hayashi, H. Ishikawa, Sensors **13**, 7053–7064 (2013)
5. M. Edmunds, R.S. Laramee, G.N. Chen, N. Max, E. Zhang, C. Ware, Comput. Graph.-Uk **36**, 974–990 (2012)
6. J.J. Lee, J.C. Dutton, A.M. Jacobi, J. Mech. Sci. Technol. **21**, 1253–1262 (2007)
7. L. Yang, H. Zare-Behtash, E. Erdem, K. Kontis, Exp. Therm. Fluid Sci. **40**, 50–56 (2012)
8. M. Schaferling, Angew. Chem. Int. Ed. **51**, 3532–3554 (2012)
9. X.D. Wang, O.S. Wolfbeis, R.J. Meier, Chem. Soc. Rev. **42**, 7834–7869 (2013)
10. J.W. Gregory, H. Sakaue, T. Liu, J.P. Sullivan, Annu. Rev. Fluid Mech. **46**, 303–330 (2014)
11. U. Fey, Y. Egami, C. Klein, ICIASF **2007**, 1–17 (2007)
12. S. Fang, S.R. Long, K.J. Disotell, J.W. Gregory, F.C. Semmelmayer, R.W. Guyton, AIAA J **50**, 109–122 (2012)
13. K. Binnemans, Chem. Rev. **109**, 4283–4374 (2009)
14. S.V. Eliseeva, J.C.G. Bunzli, Chem. Soc. Rev. **39**, 189–227 (2010)
15. J.C.G. Bünzli, S. Comby, A.S. Chauvin, C.D.B. Vandevyver, J. Rare Earths **25**, 257–274 (2007)
16. L. Armelao, S. Quici, F. Barigelletti, G. Accorsi, G. Bottaro, M. Cavazzini, E. Tondello, Coord. Chem. Rev. **254**, 487–505 (2010)
17. S. Faulkner, S.J.A. Pope, J. Am. Chem. Soc. **125**, 10526–10527 (2003)
18. S.J. Butler, D. Parker, Chem. Soc. Rev. **42**, 1652–1666 (2013)
19. R.K. Mahajan, I. Kaur, R. Kaur, S. Uchida, A. Onimaru, S. Shinoda, H. Tsukube, Chem. Commun. **17**, 2238–2239 (2003)
20. T. Gunnlaugsson, J.P. Leonard, K. Sènèchal, A.J. Harte, J. Am. Chem. Soc. **125**, 12062–12063 (2003)
21. J.-F. Lemonnier, L. Guénée, C. Beuchat, T.A. Wesolowski, P. Mukherjee, D.H. Waldeck, K.A. Gogick, S. Petoud, C. Piguet, J. Am. Chem. Soc. **133**, 16219–16234 (2011)
22. K. Miyata, T. Ohba, A. Kobayashi, M. Kato, T. Nakanishi, K. Fushimi, Y. Hasegawa, ChemPlusChem **77**, 277–280 (2012)
23. M. Mitsuishi, S. Kikuchi, T. Miyashita, Y. Amao, J. Mater. Chem. **13**, 2875–2879 (2003)
24. S.M. Borisov, O.S. Wolfbeis, Anal. Chem. **78**, 5094–5101 (2006)
25. S. Katagiri, Y. Hasegawa, Y. Wada, S. Yanagida, Chem. Lett. **33**, 1438–1439 (2004)
26. K. Miyata, Y. Konno, T. Nakanishi, A. Kobayashi, M. Kato, K. Fushimi, Y. Hasegawa, Angew. Chem. Int. Ed. **52**, 6413–6416 (2013)
27. J.F. Lemonnier, L. Guenee, C. Beuchat, T.A. Wesolowski, P. Mukherjee, D.H. Waldeck, K.A. Gogick, S. Petoud, C. Piguet, J. Am. Chem. Soc. **133**, 16219–16234 (2011)
28. A.M. Funk, P.H. Fries, P. Harvey, A.M. Kenwright, D. Parker, J. Phys. Chem. A **117**, 905–917 (2013)
29. Y. Cui, W. Zou, R. Song, J. Yu, W. Zhang, Y. Yang, G. Qian, Chem. Commun. **50**, 719–721 (2014)
30. D. Zhao, X. Rao, J. Yu, Y. Cui, Y. Yang, G. Qian, Inorg. Chem. **54**, 11193–11199 (2015)
31. X. Liu, S. Akerboom, M. de Jong, I. Mutikainen, S. Tanase, A. Meijerink, E. Bouwman, Inorg. Chem. **54**, 11323–11329 (2015)
32. H. Wang, D. Zhao, Y. Cui, Y. Yang, G. Qian. J. Solid State Chem. **246**, 341–345 (2017)
33. C. Piguet, J.C.G. Bünzli, G. Bernardinelli, G. Hopfgartner, A.F. Williams, J. Am. Chem. Soc. **115**, 8197–8206 (1993)

Chapter 4
Luminescent Lanthanide Coordination Glasses

Abstract Construction of luminescent Eu(III) coordination glasses [Eu(hfa)$_3$(*o*-dpeb)]$_2$, [Eu(hfa)$_3$(*m*-dpeb)]$_3$, and [Eu(hfa)$_3$(*p*-dpeb)]$_n$ (*o*-, *m*-, *p*-dpeb: 1,2-, 1,3-, 1,4-bis(diphenylphosphorylethynyl)benzene) are reported. The coordination structures and glass formability were dependent on the regiochemistry of substitution in regard to the internal phenylene core. Single-crystal X-ray analyses and DFT calculations revealed dimer, trimer, and polymer structures of Eu(III) coordination glasses. All of these compounds exhibited glass transition temperature in the range of 25–96 °C, and strong luminescence was also observed with intrinsic emission quantum yields above 70%.

Keywords Luminescence · Europium · Glass transition · Amorphous

4.1 Introduction

Amorphous inorganic or organic compounds play a key role in the field of materials science [1–4]. In particular, small organic molecules with amorphous formability are called "amorphous molecular materials" and expected to be applied for optical devices due to their processable, transparent, isotropic, and homogeneous properties [5, 6]. The π-conjugated organic molecules with stable amorphous state above room temperature have been developed for use in organic electroluminescent (EL) and field-effect transistor (FET) devices [7, 8]. Shirota and co-workers reported that a series of C_3-symmetrical starburst molecules such as triphenylamines and triarylbenzenes formed stable amorphous solid and exhibited charge transfer abilities (Fig. 4.1a) [9, 10]. Tian et al. demonstrated the glass formation of small organic molecules with electron-accepting 2-pyran-4-ylidenemalononitrile[11]. Balcerzak recently developed thermally stable molecular glasses composed of thiophene rings, diimide, and imine groups (T_g > 300 °C) [12].

The author here focused on Ln(III)-based luminescent glasses as a new class of amorphous molecular materials with characteristic 4f-4f transitions [13–19]. Control of molecular morphology that dominates the molecular motions and packing struc-

Fig. 4.1 a Chemical structures of small organic compounds with amorphous formability (Ar: aryl groups, X: boron, nitrogen, or benzene) and **b** design of Eu(III) coordination glasses

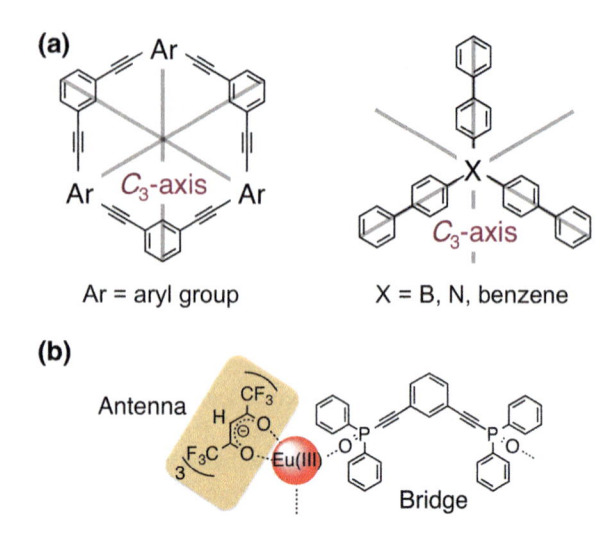

tures is required to construct amorphous Ln(III) coordination compounds. Bazan and co-workers developed an amorphous Eu(III) complex with hexyloxy groups in order to prevent easy crystal packing [20]. Long alkyl chains in Eu(III) complexes improve their solubility and amorphous formability; however, high-vibrational frequency of C–H bonds promote thermal decomposition and non-radiative quenching process of Eu(III)-centered emission [21, 22]. Therefore, novel molecular structures based on Eu(III)-β-diketonates without long alkyl chains are designed to achieve strong luminescence [23–27] (Fig. 4.1b). Hasegawa and co-workers reported the crystalline Eu(III) coordination polymers with meta- and para-substituted phenylene bridging ligands (1,3- and 1,4-bis(diphenylphosphoryl)benzene) [28, 29]. They also developed the 4,4′-substituted biphenylene bridge (4,4′-bis(diphenylphosphoryl)biphenyl: dpbp) to form rigid coordination polymers with intermolecular CH/π interactions [29, 30]. Introduction of ethynyl groups into these rigid aromatic moieties would be expected to suppress tight-binding interactions and crystallization of assembled Eu(III) complexes.

In this chapter, *ortho*-, *meta*-, and *para*-substituted phenylene bridging ligands with ethynyl groups were prepared for systematic control of the coordination structures. The morphological properties were characterized using X-ray single crystal analyses, DFT calculations, and surface observations by scanning probe microscope (SPM). The thermal and photophysical properties were evaluated based on differential scanning calorimetry (DSC) and spectroscopy.

4.2 Experimental Section

4.2.1 Materials

Europium acetate n-hydrate (99.9%), n-BuLi (in n-hexane, 1.6 M), and hydrogen peroxide were purchased from Kanto Chemical Co., Inc. 1,2-Diiodobenzene, 1,3-diethynylbenzene, 1,4-diethynylbenzene, trimethylsilylacetylene, bis(triphenylphosphine)palladium dichloride, and chlorodiphenylphosphine were obtained from Tokyo Chemical Industry Co., Ltd. All other chemicals and solvents were reagent grade and were used without further purification.

4.2.2 Apparatus

^{1}H NMR (400 MHz) spectra were recorded on a JEOL ECS400. Chemical shifts were reported in δ ppm, referenced to an internal tetramethylsilane standard for ^{1}H NMR spectroscopy. Infrared spectra were recorded on a JASCO FTIR-420 spectrometer using KBr pellets. Elemental analyses were performed by an Exeter Analytical CE440. Mass spectrometry was performed by a Thermo Scientific Exactive (ESI-MS) and a JEOL JMS-700TZ (FAB-MS).

4.2.3 Syntheses

The organic bridging ligands and corresponding coordination compounds were synthesized following Scheme 4.1.

Preparation of 1,2-bis(trimethylsilylethynyl)benzene [31]:

In a degassed 3-neck round-bottomed flask (300 mL vol.), 1,2-diiodobenzene (5.0 mL, 15 mmol), PdCl$_2$(PPh$_3$)$_2$ (0.5 g, 0.7 mmol), CuI (0.5 g, 2.5 mmol) were added and dissolved in diisopropylamine (100 mL) under argon atmosphere, then stirred until a homogeneous solution was formed at room temperature. The mixture was refluxed under argon for 4 h. The reaction mixture was allowed to cool to room temperature, ammonium salt was removed by filtration and the solvent was evaporated in vacuo. The residue was purified by column chromatography on SiO$_2$ using hexanes as an eluent to afford 1,2-bis(trimethylsilylethynyl)benzene (Yield: 3.8 g, 94%).

Preparation of 1,2-diethynylbenzene [32]:

1,2-Bis(trimethylsilylethynyl)benzene (3.8 g, 14 mmol) was dissolved in methanol (40 mL) and 1 M K$_2$CO$_3$ aqueous solution (35 mL, 35 mmol) was added. The

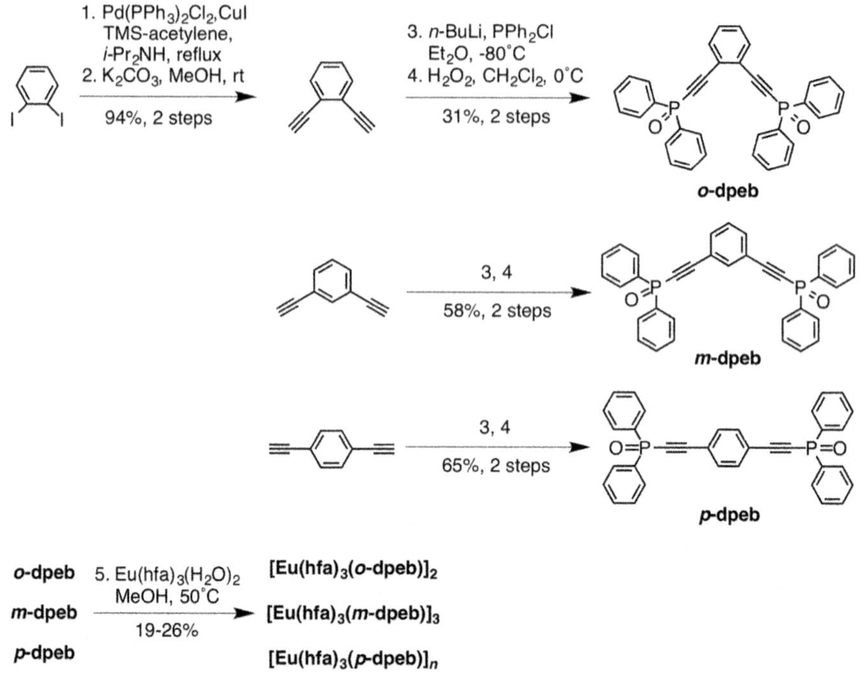

Scheme 4.1 Synthetic schemes of phosphine oxide ligands with ethynyl groups and Eu(III) coordination glasses

mixture was stirred for 3 h. The product was extracted with diethyl ether and dried over anhydrous MgSO$_4$. The solvent was evaporated to afford 1,2-diethynylbenzene as yellow oil (Yield: 1.8 g, 100%).

Preparation of 1,2-bis(diphenylphosphorylethynyl)benzene (*o*-dpeb):

In a degassed 3-neck round-bottomed flask (300 mL vol.), 1,2-diethynylbenzene (1.8 g, 14 mmol) was added and dissolved in dry diethylether (60 mL) under argon atmosphere, then stirred until a homogeneous solution was formed at room temperature. A solution of *n*-BuLi (19 mL, 31 mmol) was added dropwise to the solution at −80 °C. The addition was completed in ca. 15 min during which time the solution color changed from orange, dark green, then black. The mixture was allowed to stir for 3 h at −10 °C, after which a PPh$_2$Cl (5.6 mL, 31 mmol) was added dropwise at −80 °C, then the solution color turned to red-brown. The mixture was gradually brought to room temperature, and stirred for 14 h to form clouded brown solution. The reaction mixture was extracted with dichloromethane and dried over anhydrous MgSO$_4$. The solvent was evaporated, and the obtained pale yellow solid was placed with dichloromethane (50 mL) in a flask. The solution was cooled to 0 °C and then 30% H$_2$O$_2$ aqueous solution (34 mL) was added to it. The reaction mixture was stirred for 2 h. The product was extracted with dichloromethane, the extracts were purified by column chromatography on SiO$_2$ using ethyl acetate and hexanes as mixed eluent (ethyl acetate: hexane = 2:1).

Yield: 1.5 g (20%). ^{1}H NMR (400 MHz, CDCl$_3$, 25 °C) δ 7.79–7.86 (m, 8H, –CH), δ 7.64–7.67 (m, 2H, –CH), δ 7.43–7.49 (m, 6H, –CH), δ 7.35–7.41 (m, 8H, –CH) ppm. ESI-Mass (m/z): [M+H]$^+$ calcd for C$_{34}$H$_{25}$O$_2$P$_2$, 527.1; found, 527.1. Anal. Calcd for C$_{34}$H$_{24}$O$_2$P$_2$: C, 75.56; H, 4.59%. Found: C, 75.99; H, 4.58%.

Preparation of 1,3-bis(diphenylphosphorylethynyl)benzene (*m*-dpeb):

The 1,3-diethynylbenzene was dissolved in dry diethylether (50 mL) and the mixture degassed by argon bubbling for 20 min. A solution of *n*-BuLi (13 mL, 21 mmol) was added dropwise to the solution at −80 °C. The addition was completed in ca. 15 min during which time a yellow precipitate was formed. The mixture was allowed to stir for 3 h at −10 °C, after which a PPh$_2$Cl (3.8 mL, 21 mmol) was added dropwise at −80 °C. The mixture was gradually brought to room temperature, and stirred for 14 h. The product was extracted with dichloromethane, the extracts washed with brine for three times and dried over anhydrous MgSO$_4$. The solvent was evaporated, and the obtained pale yellow solid was placed with dichloromethane (50 mL) in a flask. The solution was cooled to 0 °C and then 30% H$_2$O$_2$ aqueous solution (10 mL) was added to it. The reaction mixture was stirred for 2 h. The product was extracted with dichloromethane and purified by column chromatography on SiO$_2$ using ethyl acetate and hexanes as mixed eluent (ethyl acetate: hexane = 2:1).

Yield: 4.3 g (58%). ^{1}H–NMR (400 MHz, CDCl$_3$, 25 °C) δ 7.84–7.95 (m, 8H, –CH), δ 7.82 (s, 1H, –CH) δ 7.67–7.70 (d, 1H, –CH), δ 7.64–7.66 (d, 1H, –CH), δ 7.47–7.62 (m, 12H, –CH), δ 7.38–7.46 (t, 1H, –CH) ppm. ESI-Mass (m/z): [M+H]$^+$ calcd for C$_{34}$H$_{25}$O$_2$P$_2$, 527.1; found, 527.1. Anal. Calcd for C$_{34}$H$_{24}$O$_2$P$_2$: C, 77.56; H, 4.59%. Found: C, 77.70; H, 4.64%.

Preparation of 1,4-bis(diphenylphosphorylethynyl)benzene (*p*-dpeb):

The 1,4-diethynylbenzene was dissolved in dry diethylether (50 mL) and the mixture degassed by Ar bubbling for 20 min. A solution of *n*-BuLi (13 mL, 21 mmol) was added dropwise to the solution at −80 °C. The addition was completed in ca. 15 min during which time a yellow precipitate was formed. The mixture was allowed to stir for 3 h at −10 °C, after which a PPh$_2$Cl (3.8 mL, 21 mmol) was added dropwise at −80 °C. The mixture was gradually brought to room temperature, and stirred for 14 h. The product was extracted with dichloromethane, the extracts washed with brine for three times and dried over anhydrous MgSO$_4$. The solvent was evaporated, and the obtained pale yellow solid was placed with dichloromethane (50 mL) in a flask. The solution was cooled to 0 °C and then 30% H$_2$O$_2$ aqueous solution (10 mL) was added to it. The reaction mixture was stirred for 2 h. The product was extracted with dichloromethane and purified by column chromatography on SiO$_2$ using ethyl acetate and hexanes as mixed eluent (ethyl acetate: hexane = 2:1).

Yield: 4.3 g (58%). ^{1}H NMR (400 MHz, CDCl$_3$, 25 °C) δ 7.84–7.90 (m, 8H, –CH), δ 7.59 (s, 4H, –CH), δ 7.55–7.56 (d, 4H, –CH), δ 7.47–7.51 (m, 8H, –CH) ppm. ESI-Mass (m/z): [M+H]$^+$ calcd for C$_{34}$H$_{25}$O$_2$P$_2$, 527.1; found, 527.1. Anal. Calcd for C$_{34}$H$_{24}$O$_2$P$_2$: C, 77.56; H, 4.59%. Found: C, 77.52; H, 4.72%.

Preparation of Eu(III) complexes [Eu(hfa)₃(o-dpeb)]₂ and [Eu(hfa)₃(p-dpeb)]ₙ:

The phosphine oxide ligands (0.21 g, 0.40 mmol) and Eu(hfa)₃(H₂O)₂ (0.32 g, 0.40 mmol) were dissolved in methanol (20 mL), respectively. The solutions were mixed and stirred for 2 h. The solvent was evaporated, and the obtained white solid was washed with chloroform and hexanes.

[Eu(hfa)₃(o-dpeb)]₂: Yield 0.27 g (26%). IR (KBr) 1657 (st, C=O), 1132 (st, P=O), 1093–1249 (st, C–O–C and st, C–F) cm⁻¹. FAB-Mass (m/z): [M-hfa]⁺ calcd for C₉₃H₅₃Eu₂F₃₀O₁₄P₄, 2392.2; found, 2392.0. Anal. Calcd for: C, 45.29; H, 2.09%. Found: C, 45.05; H, 2.21%.

[Eu(hfa)₃(p-dpeb)]ₙ: Yield 0.39 g (75% for monomer). IR (KBr) 1657 (st, C=O), 1141 (st, P=O), 1093–1249 (st, C–O–C and st, C–F) cm⁻¹. FAB-Mass (m/z): [M-hfa]⁺ calcd for C₄₄H₂₆EuF₁₂O₆P₂, 1092.6; found, 1092.7. Anal. Calcd for: C, 45.29; H, 2.09%. Found: C, 44.98; H, 2.18%.

Preparation of an Eu(III) complex [Eu(hfa)₃(m-dpeb)]₃:

The m-dpeb ligand (0.21 g, 0.40 mmol) and Eu(hfa)₃(H₂O)₂ (0.32 g, 0.40 mmol) were dissolved in methanol (20 mL), respectively. The solutions were mixed and stirred for 3 h at 50 °C. The reaction mixture was concentrated, re-dissolved in methanol (5 mL), and then hexane (15 mL) was added. The organic solvents were removed by decompression to form amorphous solids.

Yield: 0.34 g (22%). IR (KBr) 1657 (st, C=O), 1142 (st, P=O), 1093–1258 (st, C–O–C and st, C–F) cm⁻¹. FAB-Mass (m/z): [M-hfa]⁺ calcd for C₁₄₂H₈₀Eu₃F₄₈O₂₂P₆, 3691.8; found, 3691.8. Anal. Calcd for: C, 45.29; H, 2.09%. Found: C, 45.48; H, 2.51%

4.2.4 Crystallography

Colorless single crystals of the complexes were mounted on glass fibers using paraffin oil. All measurements were made on a Rigaku RAXIS RAPID imaging plate area detector with graphite monochromated MoK$_\alpha$ radiation. Correction for decay and Lorentz-polarization effects were made using empirical absorption correction, solved by direct methods and expanded using Fourier techniques. Non-hydrogen atoms were refined anisotropically. Hydrogen atoms were refined using the riding model. The final cycle of full-matrix least-squares refinement was based on observed reflections and variable parameters. All calculations were performed using the crystal structure crystallographic software package. The CIF data was confirmed by the checkCIF/PLATON service. CCDC-1037529 (for [Eu(hfa)₃(o-dpeb)]₂) and CCDC 1037531 (for [Eu(hfa)₃(p-dpeb)]ₙ) contain the supplementary crystallographic data for this paper. These data can be obtained free of charge from The Cambridge Crystallographic Data Centre via www.ccdc.cam.ac.uk/data_request/cif.

4.2 Experimental Section

4.2.5 Optical Measurements

UV-Vis absorption spectra were recorded on a JASCO V-670 spectrometer. Emission and excitation spectra were recorded on a HORIBA Fluorolog-3 spectrofluorometer and corrected for the response of the detector system. Emission lifetimes (τ_{obs}) were measured using the third harmonics (355 nm) of a Q-switched Nd:YAG laser (Spectra Physics, INDI-50, fwhm = 5 ns, λ = 1064 nm) and a photomultiplier (Hamamatsu photonics, R5108, response time $\leq$ 1.1 ns). The Nd:YAG laser response was monitored with a digital oscilloscope (Sony Tektronix, TDS3052, 500 MHz) synchronized to the single-pulse excitation. Emission lifetimes were determined from the slope of logarithmic plots of the decay profiles. The emission quantum yields excited at 380 nm (Φ_{tot}) were estimated using JASCO F-6300-H spectrometer attached with JASCO ILF-533 integrating sphere unit (φ = 100 mm). The wavelength dependence of the detector response and the beam intensity of Xe light source for each spectrum were calibrated using a standard light source.

4.2.6 DFT Calculations

Due to lack in the structural data of [Eu(hfa)(*m*-dpeb)]$_3$, the DFT calculation was carried out at B3LYP level with the basis sets SDD for Eu and 6-31G** for other atoms using the Gaussian 09 program package [33–36]. The ultrafine grid was used for numerical integrations. A trimer structure was plausibly assumed for this complex.

4.3 Results and Discussion

4.3.1 Structural Characterization

Single crystals were prepared for [Eu(hfa)$_3$(*o*-dpeb)]$_2$ and [Eu(hfa)$_3$(*p*-dpeb)]$_n$. Crystallographic data of these compounds were summarized in Table 4.1. The Eu(III) ions for both compounds formed typical 8-coordination with three hfa and two bridging ligands. As described in Chap. 2, the coordination geometry around Eu(III) ions was examined on the basis of shape factor *S*. The observed dihedral angles (δ_i), ideal dihedral angles for square antiprism (θ_{SAP}) and trigonal dodecahedron (θ_{TDH}), calculated measure shape criteria, S_{SAP} and S_{TDH} were summarized (Figs. 4.2 and 4.3; Tables 4.2 and 4.3). The *S* values of [Eu(hfa)$_3$(*o*-dpeb)]$_2$ and [Eu(hfa)$_3$(*p*-dpeb)]$_n$ for 8-SAP structures (S_{SAP} = 6.90 and 5.17, respectively) were smaller than those for 8-TDH (S_{TDH} = 11.1 and 13.5, respectively), suggesting that the coordination geometry of these compounds was categorized to be 8-SAP. The author also focused on assembled structures of these compounds. [Eu(hfa)$_3$(*o*-dpeb)]$_2$ formed a C_2-symmetrical dimer structure composed of two Eu(III) ions, six hfa, and two phosphine oxide bridg-

Table 4.1 Crystallographic data of Eu(III) coordination compounds

	[Eu(hfa)$_3$(o-dpeb)]$_2$	[Eu(hfa)$_3$(p-dpeb)]$_n$
Chemical formula	C$_{98}$H$_{54}$Eu$_2$F$_{36}$O$_{16}$P$_4$	C$_{49}$H$_{27}$EuF$_{18}$O$_8$P$_2$
Formula weight	2599.25	1299.63
Crystal system	Monoclinic	Monoclinic
Space group	$P2_1/n$ (#14)	$P2_1/c$ (#14)
a/Å	12.5432(9)	23.1941(10)
b/Å	30.3644(15)	13.4114(5)
c/Å	12.9212(9)	16.7226(6)
α/deg	90.000	90.000
β/deg	90.378(6)	91.6148(10)
γ/deg	90.000	90.000
Volume/Å^3	4921.1(5)	5199.8(4)
Z	2	4
d_{calc}/g cm^{-3}	1.754	1.660
Temperature/ °C	23.0	−150
μ (Mo K$_\alpha$)/cm^{-1}	14.590	13.809
max 2θ/deg	55.0	55.00
Reflections collected	48310	49124
Independent reflections	11257	11871
R_1	0.0439	0.0420
wR_2	0.0728	0.0998

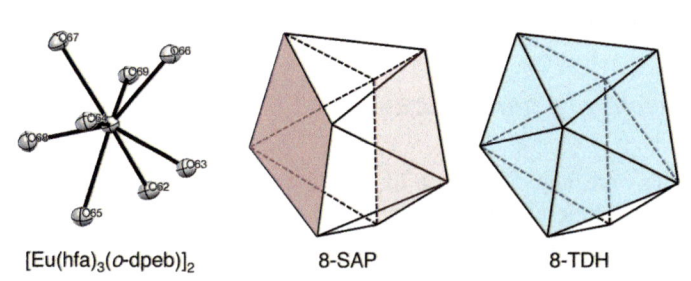

$^a R_1 = \sum ||F_o| - |F_c|| / \sum F_o$, $^b wR_2 = [\sum w (F_o^2 - F_c^2)^2 / \sum w (F_o^2)^2]^{1/2}$

Fig. 4.2 Coordination environments around an Eu(III) ion of [Eu(hfa)$_3$(o-dpeb)]$_2$

ing ligands with multiple intra and intermolecular CH/F ($d_{CH/F}$ = 2.63 Å (H39/F6), 3.00 Å (H44/F4), and 2.79 Å (H56/F9)) and π/π ($d_{\pi/\pi}$ = 3.19 and 3.43 Å) interactions (Fig. 4.4). A large crystal density (1.754 g cm^{-3}) was responsible for the highly-ordered packing structure.

[Eu(hfa)$_3$(p-dpeb)]$_n$ formed a polymer chain (Fig. 4.5) with intra and intermolecular CH/F interactions ($d_{CH/F}$ = 3.03 Å (H25/F73), 2.63 Å (H46/F38), and 2.69 Å

4.3 Results and Discussion

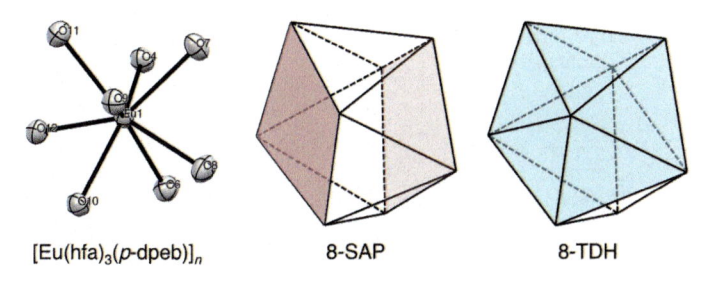

Fig. 4.3 Coordination environments around an Eu(III) ion of [Eu(hfa)$_3$(p-dpeb)]$_n$

Table 4.2 The observed (δ_i), idealized dihedral angles (θ_{SAP}, θ_{TDH}), and calculated S values ($S_{8\text{-SAP}}$, $S_{8\text{-TDH}}$) for [Eu(hfa)$_3$(o-dpeb)]$_2$

	δ_i	SAP θ_{SAP}	$\delta_i - \theta_{SAP}$	$(\delta_i - \theta_{SAP})^2$	TDH θ_{TDH}	$\delta_i - \theta_{TDH}$	$(\delta_i - \theta_{TDH})^2$
O64–O68	17.52	0	17.52	306.95	29.86	−12.34	152.28
O64–O67	69.82	77.1	−7.28	53.00	61.48	8.34	69.56
O64–O65	69.04	77.1	−8.06	64.96	74.29	−5.25	27.56
O68–O67	71.43	77.1	−5.67	32.15	74.29	−2.86	8.18
O68–O65	71.85	77.1	−5.25	27.56	61.48	10.37	107.54
O63–O69	0.88	0	0.88	0.77	29.86	−28.98	839.84
O63–O62	74.61	77.1	−2.49	6.20	61.48	13.13	172.40
O63–O66	80.38	77.1	3.28	10.76	74.29	6.09	37.09
O69–O66	72.54	77.1	−4.56	20.79	76.62	−4.08	16.65
O69–O62	84.73	77.1	7.63	58.22	74.84	9.89	97.81
O62–O68	43.07	51.6	−8.53	72.76	54.51	−11.44	130.87
O62–O65	58.75	51.6	7.15	51.12	48.09	10.66	113.64
O63–O64	47.73	51.6	−3.87	14.98	47.76	−0.03	0.0009
O63–O65	51.74	51.6	0.14	0.02	54.48	−2.74	7.51
O66–O64	44.26	51.6	−7.34	53.88	52.57	−8.31	69.06
O66–O67	60.19	51.6	8.59	73.79	52.85	7.34	53.88
O69–O68	53.04	51.6	1.44	2.07	47.34	5.70	32.49
O69–O67	54.25	51.6	2.65	7.02	52.63	1.62	2.62
		$S_{8\text{-SAP}} = 6.90$			$S_{8\text{-TDH}} = 10.38$		

(H22/F38)). [Eu(hfa)$_3$(p-dpeb)]$_n$ showed relatively loose-packed structure due to the absence of intermolecular CH/π interactions between single polymer chains, unlike reported rigid Eu(III) coordination polymer, [Eu(hfa)$_3$(dpbp)]$_n$ [29]. The author considered this to be responsible for ethynyl groups in p-dpeb that provided a large crystal volume with small availability of aromatic surfaces for intermolecular CH/π interactions. On the other hand, [Eu(hfa)$_3$(m-dpeb)]$_3$ formed amorphous solid. A trimer structure was assumed based on the fragments of Eu$_3$(hfa)$_8$(m-dpeb)$^{3+}$ in FAB-MS spectrum (Fig. 4.6a). The X-ray diffraction image and patterns also indicated the

Table 4.3 The observed (δ_i), idealized dihedral angles (θ_{SAP}, θ_{TDH}), and calculated S values ($S_{8\text{-SAP}}$, $S_{8\text{-TDH}}$) for [Eu(hfa)$_3$(p-dpeb)]$_n$

		SAP			TDH		
	δ_i	θ_{SAP}	$\delta_i - \theta_{SAP}$	$(\delta_i - \theta_{SAP})^2$	θ_{TDH}	$\delta_i - \theta_{TDH}$	$(\delta_i - \theta_{TDH})^2$
O9–O12	10.03	0	10.03	100.60	29.86	−19.83	393.23
O9–O11	82.39	77.1	5.29	27.98	61.48	20.91	437.23
O9–O10	75.35	77.1	−1.75	3.06	74.29	1.06	1.12
O12–O11	83.53	77.1	6.43	41.34	74.29	9.24	85.38
O12–O10	83.40	77.1	6.3	39.69	61.48	21.92	480.49
O8–O4	10.15	0	10.15	103.02	29.86	−19.71	388.48
O8–O6	84.85	77.1	7.75	60.06	61.48	23.37	546.16
O8–O7	78.25	77.1	1.15	1.32	74.29	3.96	15.68
O4–O7	81.15	77.1	4.05	16.40	76.62	4.53	20.52
O4–O6	78.17	77.1	1.07	1.14	74.84	3.33	11.09
O6–O12	51.51	51.6	−0.09	0.0081	54.51	−3.00	9.00
O6–O10	49.56	51.6	−2.04	4.16	48.09	1.47	2.16
O8–O9	55.22	51.6	3.62	13.10	47.76	7.46	55.65
O8–O10	48.86	51.6	−2.74	7.51	54.48	−5.62	31.58
O7–O9	53.79	51.6	2.19	4.80	52.57	1.22	1.49
O7–O11	45.69	51.6	−5.91	34.93	52.85	−7.16	51.27
O4–O12	55.93	51.6	4.33	18.75	47.34	8.59	73.79
O4–O11	51.08	51.6	−0.52	0.27	52.63	−1.55	2.40
		$S_{8\text{-SAP}} = 5.15$			$S_{8\text{-TDH}} = 12.03$		

lack of crystallinity and the obtained signals were similar to those of glass ceramics (Fig. 4.6b, c) [37].

4.3.2 DFT Optimization

Optimization of [Eu(hfa)$_3$(o-dpeb)]$_2$ was first carried out. The dimer structure determined by single crystal X-ray analysis was adopted as the initial structure for DFT calculation. The optimized structure was well corresponded with that of X-ray analysis (Fig. 4.7). The unit cell of [Eu(hfa)$_3$(p-dpeb)]$_n$ was also calculated as a closed-shell singlet configuration since it was composed of an Eu(III) ion and three anionic hfa ligands. The optimized structure agreed with that of single crystal X-ray crystallography (Fig. 4.8). According to the flat band structure of [Eu(hfa)$_3$(p-dpeb)]$_n$, conjugation was not extended over the entire polymer chains (band gap at Γ and X points were estimated to be 4.0420 and 4.0231 eV, respectively). The estimated band gap energy ($\approx$4.0 eV) for [Eu(hfa)$_3$(o-dpeb)]$_2$ and [Eu(hfa)$_3$(p-dpeb)]$_n$ matched the S_0-S_1 gap of hfa ligands.

Based on these results, optimization of [Eu(hfa)$_3$(m-dpeb)]$_3$ was carried out using the same basis sets as those for the other two compounds. The initial structure was

4.3 Results and Discussion 57

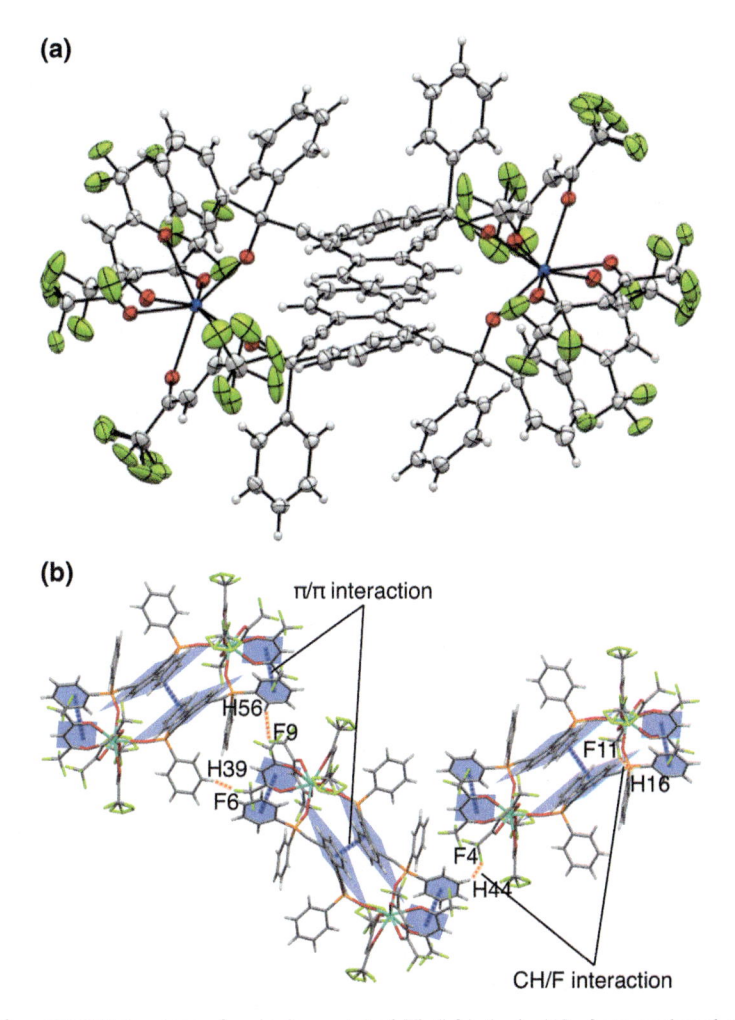

Fig. 4.4 a ORTEP drawings of a single crystal of [Eu(hfa)$_3$(*o*-dpeb)]$_2$ from methanol solution. Thermal ellipsoids are shown at 50% probability level. **b** Capped sticks model of [Eu(hfa)$_3$(*o*-dpeb)]$_2$ focused on the selected intra and intermolecular interactions. Dashed lines denote CH/F and π/π interactions

determined to be the assumed trimer structure, and the *pseudo*-C_3 symmetrical moiety was optimized (Fig. 4.9). The [Eu(hfa)$_3$(*m*-dpeb)]$_3$ showed highly broken symmetry compared to that of the C_2-symmetrical [Eu(hfa)$_3$(*o*-dpeb)]$_2$. Multiple metastable states observed during the calculations were probably related to the broken symmetry of [Eu(hfa)$_3$(*m*-dpeb)]$_3$ and contributed to the formation of an amorphous state instead of crystal packing.

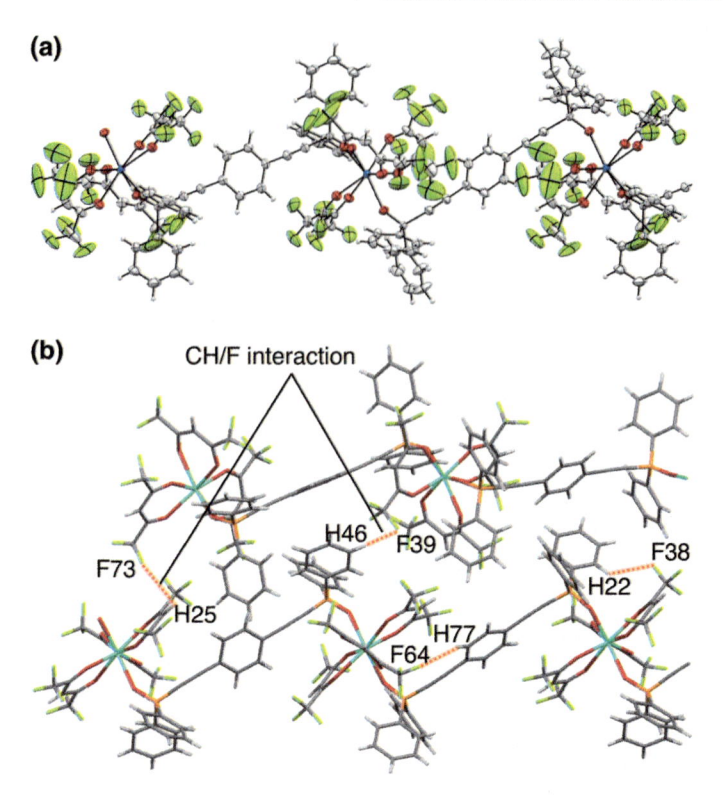

Fig. 4.5 a ORTEP drawings of a single crystal of [Eu(hfa)$_3$(*p*-dpeb)]$_n$ from methanol solution. Thermal ellipsoids are shown at 50% probability level. **b** Capped sticks model of [Eu(hfa)$_3$(*p*-dpeb)]$_n$ focused on the selected intra and intermolecular interactions. Dashed lines denote CH/F interactions

4.3.3 Thermal Properties

DSC measurements were conducted to evaluate the thermophysical properties of the Eu(III) complexes (Fig. 4.10). [Eu(hfa)$_3$(*o*-dpeb)]$_2$ exhibited a glass transition point (T_g) at 25 °C by cooling the supercooled liquid, which is a characteristic of amorphous molecular materials. Structural relaxation of amorphous molecular materials generally occurs from glass to crystal via a meta-stable supercooled liquid state. In the presence of a polymorph, these compounds finally form a crystal with the highest melting point based on the solid-solid phase transitions or melt-crystallization [38, 39]. Here the structural rearrangement of dimer compounds were frozen below T_g to form a glassy state. [Eu(hfa)$_3$(*p*-dpeb)]$_n$ also showed T_g at 46 °C, corresponding to the relaxation of segmental movements of polymer chains. The author considered a broad endothermic peak at 96 °C to be responsible for the melting point of various-sized crystals.

4.3 Results and Discussion

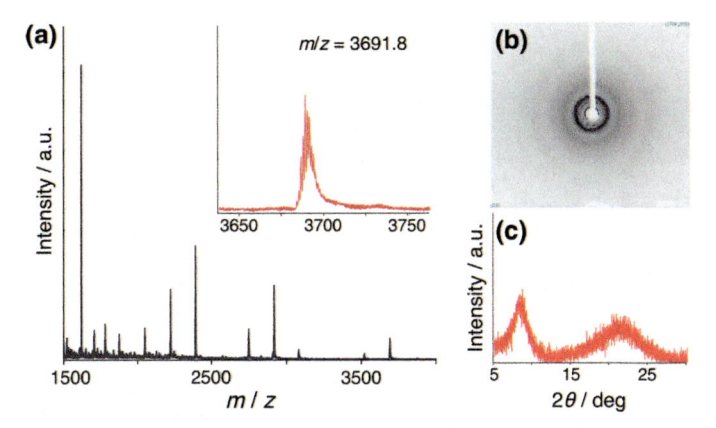

Fig. 4.6 **a** FAB-MS spectrum, **b** X-ray diffraction image and **c** powder X-ray diffraction patterns of [Eu(hfa)$_3$(*m*-dpeb)]$_3$

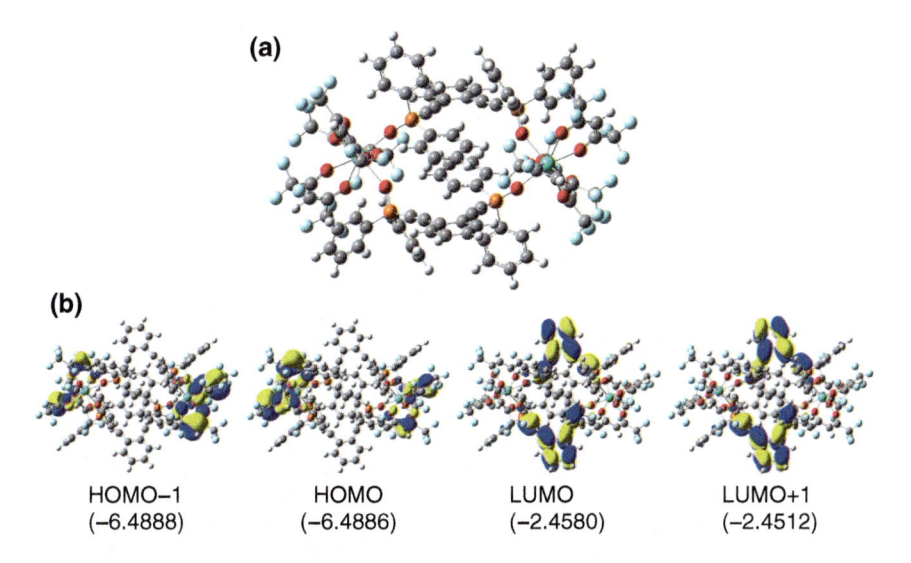

Fig. 4.7 **a** DFT optimized structure and **b** HOMO-LUMO frontier orbitals of [Eu(hfa)$_3$(*o*-dpeb)]$_2$

The T_g of [Eu(hfa)$_3$(*m*-dpeb)]$_3$ was identified at 65 °C. In the case of [Eu(hfa)$_3$(*m*-dpeb)]$_3$, crystals are not formed above T_g, since the most thermodynamically stable structure has highly broken symmetry as shown in the DFT optimization and crystallization rate is expected to be low. A sharp endothermic peak was observed at around T_g in first heating process, and a viscous liquid was formed above this temperature. These thermodynamic behaviors are similar to that of complete amorphous polymer [40]. Thus, the sharp endothermic peak around T_g in the first run is probably responsible for the enthalpy relaxation of [Eu(hfa)$_3$(*m*-dpeb)]$_3$ due to the stable-glass formation.

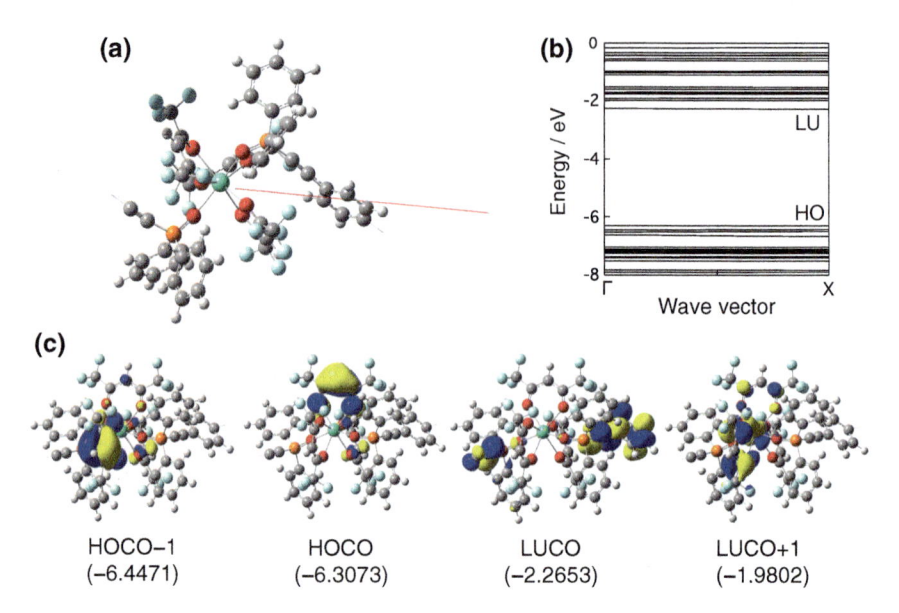

Fig. 4.8 **a** DFT optimized structure, **b** band structure, and **c** HOCO-LUCO structures at the Γ point (HOCO: highest occupied crystal orbital, LUCO: lowest unoccupied crystal orbital) of [Eu(hfa)$_3$(*p*-dpeb)]$_n$

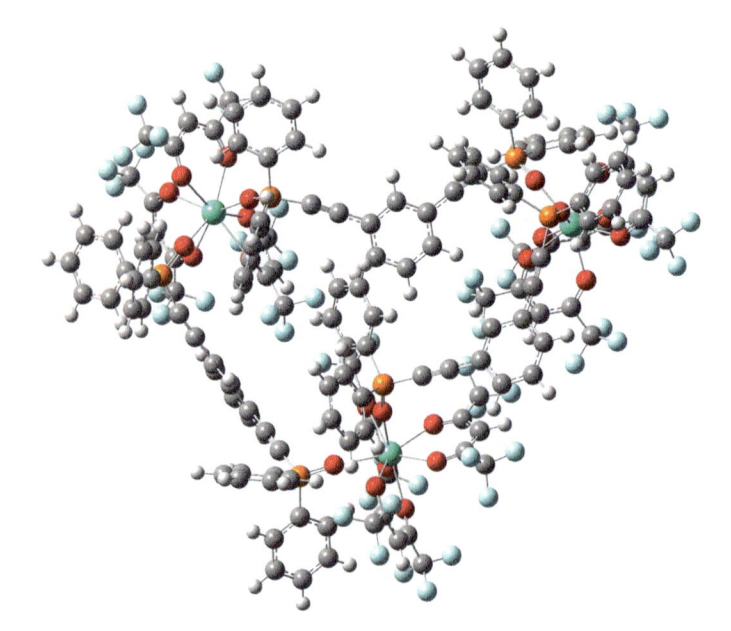

Fig. 4.9 DFT optimized structure of [Eu(hfa)$_3$(*m*-dpeb)]$_3$

4.3 Results and Discussion

Fig. 4.10 DSC thermograms of **a** [Eu(hfa)$_3$(o-dpeb)]$_2$, **b** [Eu(hfa)$_3$(m-dpeb)]$_3$, **c** [Eu(hfa)$_3$(p-dpeb)]$_n$ and chemical structures of bridging ligands (argon atmosphere, heating/cooling rate = 5 °C min^{-1}, dashed line: cooling process, solid line: heating process)

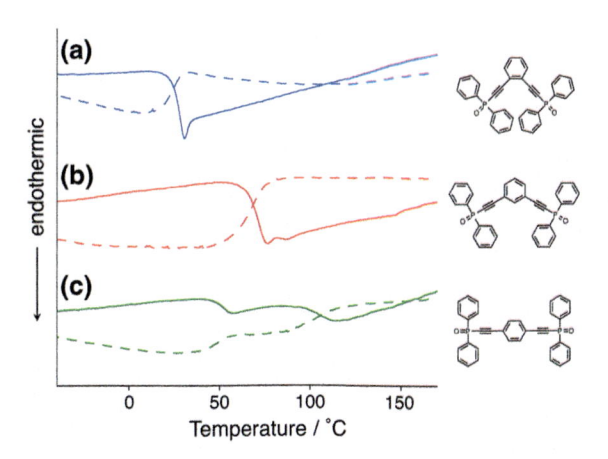

These results indicate that the bent angles of the dpeb ligands (60°, 120°, 180°) have an influence on the molecular morphologies and the glass-transition properties. The author also considers that these glass-transition phenomena are caused by the presence of ethynyl groups in the bent-angled dpeb ligands, although phenylene-bridged Eu(III) coordination polymers with tight-packing structures have no glass-transition points [28, 29].

4.3.4 Surface Observations by Scanning Probe Microscope (SPM)

The crystal surfaces of obtained compounds were observed using a scanning probe microscope. A step-like surface was observed for [Eu(hfa)$_3$(o-dpeb)]$_2$ (Fig. 4.11a). The step heights were estimated to be 1.56 or 3.13 nm, which corresponded to the size of one or two [Eu(hfa)$_3$(o-dpeb)]$_2$ molecules (1.52 and 3.05 nm, respectively). [Eu(hfa)$_3$(p-dpeb)]$_n$ formed a fibrous surface structure due to the polymer chains in the crystal packing system (Fig. 4.11c). The height profile was also analyzed and obtained values (0.99 to 1.01 nm) almost matched the width of a single polymer chain (1.08 nm). On the contrary to the crystalline [Eu(hfa)$_3$(o-dpeb)]$_2$ and [Eu(hfa)$_3$(p-dpeb)]$_n$, a flat and smooth surface was observed for [Eu(hfa)$_3$(m-dpeb)]$_3$ (Fig. 4.11b). Step- or fiber-like terrace structures were not found, and the height profiles indicated that no regularity is identified due to the amorphous property. The crystal structures and molecular shapes are highly reflected on the SPM images, and molecular arrangement can affect on entire solid-state structures.

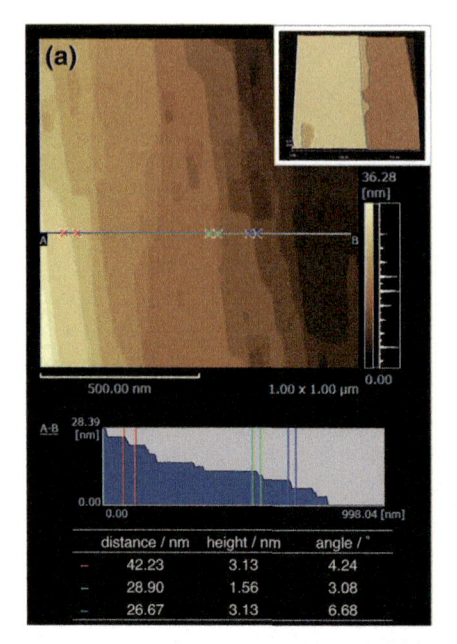

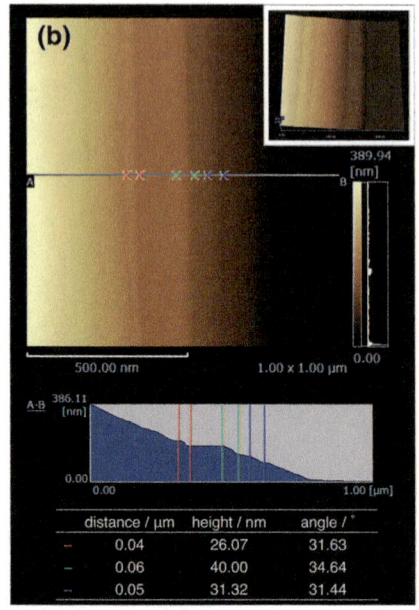

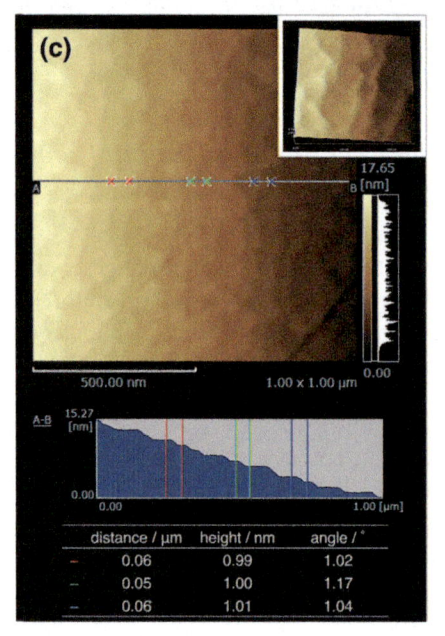

Fig. 4.11 1 μm × 1 μm SPM images and height profiles of **a** [Eu(hfa)$_3$(*o*-dpeb)]$_2$, **b** [Eu(hfa)$_3$(*m*-dpeb)]$_3$, and **c** [Eu(hfa)$_3$(*p*-dpeb)]$_n$ (inset shows 250 × 250 nm image)

4.3.5 Photophysical Properties

The emission spectra in the solid state are shown in Fig. 4.12a. Emission bands observed at 578, 591, 613, 650, and 698 nm are attributed to the 4f-4f transitions of Eu(III) ions ($^5D_0 \rightarrow {}^7F_J$: $J = 0, 1, 2, 3, 4$). The stark splitting energy of $^5D_0 \rightarrow {}^7F_2$ is generally depended on the coordination geometry [28, 41]. The stark splitting was not clearly observed for [Eu(hfa)$_3$(*m*-dpeb)]$_3$, compared to [Eu(hfa)$_3$(*o*-dpeb)]$_2$ and [Eu(hfa)$_3$(*p*-dpeb)]$_n$. This can be explained by the isotropy and low crystallinity of amorphous compounds. The time-resolved emission profiles for these compounds revealed single-exponential decay with millisecond scale lifetimes (Fig. 4.12c). The photophysical parameters such as intrinsic emission quantum yields (Φ_{ff}), radiative and nonradiative rate constants (k_r, k_{nr}) were estimated using the Eqs. (2.3–2.5) in Chap. 2 and summarized in Table 4.4.

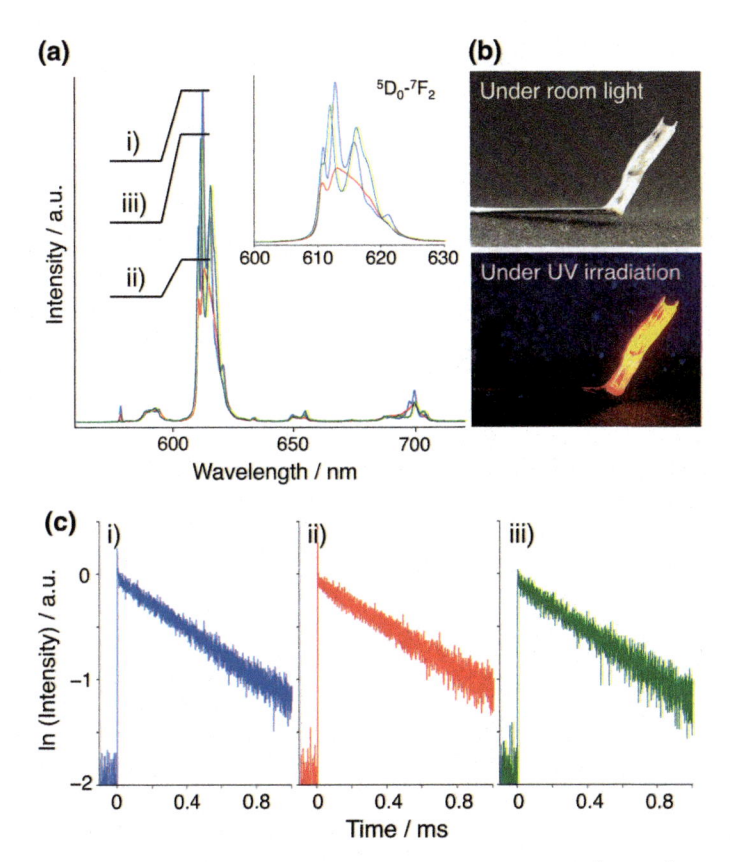

Fig. 4.12 **a** Emission spectra in solid state ($\lambda_{ex} = 465$ nm, inset shows $^5D_0 \rightarrow {}^7F_2$ transition), **b** pictures of amorphous solid under room light and under UV-irradiation, and **c** emission decay profiles of (i) [Eu(hfa)$_3$(*o*-dpeb)]$_2$, (ii) [Eu(hfa)$_3$(*m*-dpeb)]$_3$, and (iii) [Eu(hfa)$_3$(*p*-dpeb)]$_n$

Table 4.4 Photophysical parameters of Eu(III) coordination glasses in solid state

Compounds	$\Phi_{tot}{}^a$ (%)	$\Phi_{ff}{}^b$ (%)	$\tau_{obs}{}^c$ (ms)	$k_r{}^b$ (s^{-1})	$k_{nr}{}^b$ (s^{-1})
[Eu(hfa)$_3$(o-dpeb)]$_2$	52	94	1.0	9.1×10^2	6.1×10
[Eu(hfa)$_3$(m-dpeb)]$_3$	44	72	1.0	7.2×10^2	2.8×10^2
[Eu(hfa)$_3$(p-dpeb)]$_n$	54	86	0.93	9.2×10^2	1.5×10^2

$^a\lambda_{ex} = 380$ nm, ^{b}Equations (2.3–2.5) in Chap. 2, $^c\lambda_{ex} = 355$ nm

The [Eu(hfa)$_3$(o-dpeb)]$_2$ showed large Φ_{ff} (94%) due to the large k_r and small k_{nr} values. The large k_r for [Eu(hfa)$_3$(o-dpeb)]$_2$ (9.1×10^2 s^{-1}) and [Eu(hfa)$_3$(p-dpeb)]$_n$ (9.2×10^2 s^{-1}) reflect the enhancement of electric dipole transition probability under the asymmetrical coordination geometries. The significantly small k_{nr} for [Eu(hfa)$_3$(o-dpeb)]$_2$ is induced by the multiple intramolecular CH/F and π/π interactions in the dimer structures. The relatively small k_r (7.2×10^2 s^{-1}) and large k_{nr} (2.8×10^2 s^{-1}) of [Eu(hfa)$_3$(m-dpeb)]$_3$ were due to the isotropic and loose-packed character of amorphous solid. By taking advantage of hard-to-crystallize property, the notable amorphous form of luminescent [Eu(hfa)$_3$(m-dpeb)]$_3$ without polymer matrices was demonstrated in Fig. 4.12b.

The transmittance spectrum of the glassy-state [Eu(hfa)$_3$(m-dpeb)]$_3$ in a laminate film also supports the transparency (~70%) in visible region (Fig. 4.13a). The small spike observed at 465 nm is 4f-4f absorption of Eu(III) ions. The bright red luminescence was observed under UV irradiation. When conventional Eu(III) complexes with β-diketonates (absorption coefficient > 30,000 L mol^{-1} cm^{-1}) are embedded in polymer thin films (thickness: 0.1 μm, dopant concentration: 1 wt%), the absorbance of the films are estimated to be approximately 0.003. On the other hand, absorbance of thin films prepared from Eu(III) coordination glasses is estimated to be 0.3. These simple estimations for absorbance indicate that the latter film would be much brighter than those of former films. The transparent films without scattering from crystal grain boundaries on glass substrate (Fig. 4.13b, c) also indicate the excellent coating ability of luminescent Eu(III) coordination glass.

4.4 Conclusions

In conclusion, the author successfully controlled the morphological properties of novel luminescent Eu(III) coordination compounds by introducing *ortho*-, *meta*-, and *para*-substituted phenylene bridges with ethynyl groups. The coordination structures were determined to be dimer, trimer, and polymer structures, respectively. All of these compounds exhibited glass transition points, and the trimer compound with 120°-angled *meta*-substituted bridges formed stable amorphous state. The glass transition properties were dependent on the regiochemistry of substitution in regard to the internal phenylene core. They exhibited relatively strong emission as well.

4.4 Conclusions

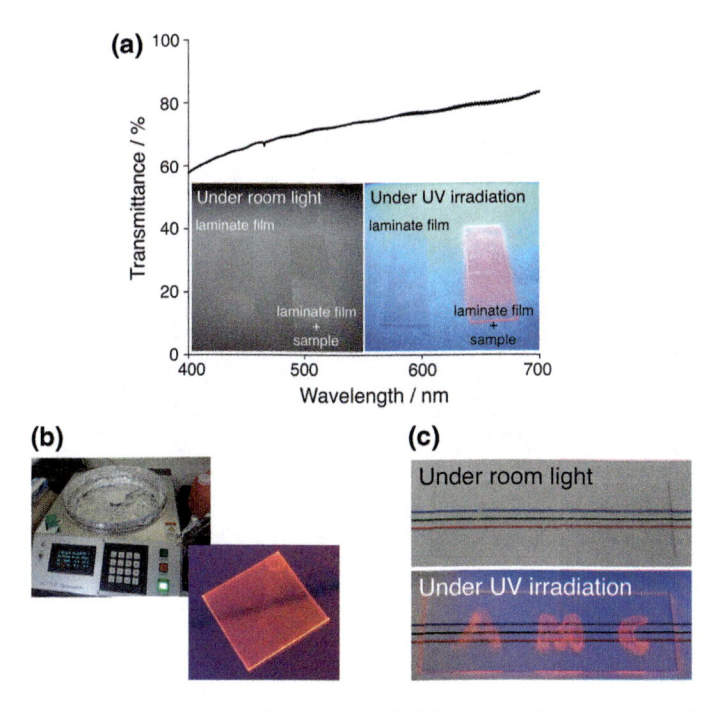

Fig. 4.13 a Transmittance spectrum of glassy-state [Eu(hfa)$_3$(m-dpeb)]$_3$ in a laminate film (pictures at the bottom show the laminate films with and without samples under room light and UV). **b** Spin-coated film and **c** amorphous painting on glass substrates

These results would provide a guideline on molecular designs to control the entire structures and thermal properties of assembled Ln(III) complexes via small organic bridging ligands. Ln(III) complexes with amorphous-forming and strong-luminescence properties that can be processed at ambient temperatures are also expected as paintable luminescent materials without polymer matrices.

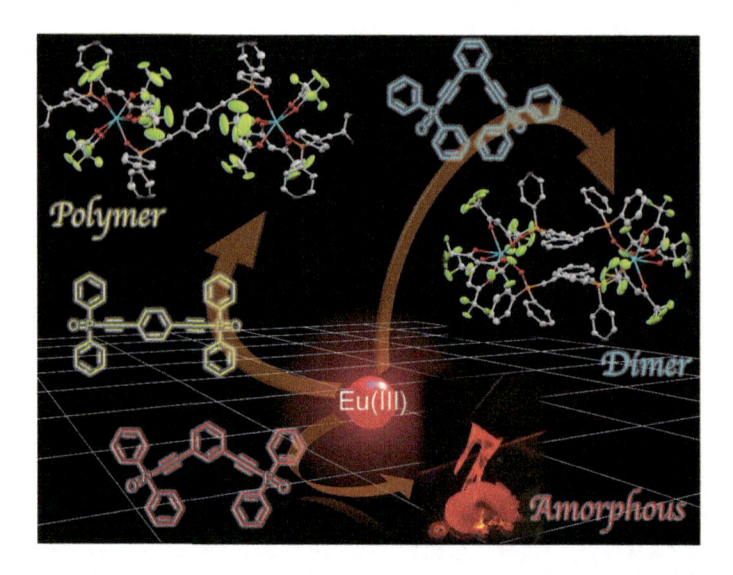

References

1. S.J. Wang, W.J. Oldham, R.A. Hudack, G.C. Bazan, J. Am. Chem. Soc. **122**, 5695–5709 (2000)
2. S.Y. Reece, J.A. Hamel, K. Sung, T.D. Jarvi, A.J. Esswein, J.J.H. Pijpers, D.G. Nocera, Sci. **334**, 645–648 (2011)
3. R.D.L. Smith, M.S. Prevot, R.D. Fagan, Z.P. Zhang, P.A. Sedach, M.K.J. Siu, S. Trudel, C.P. Berlinguette, Sci. **340**, 60–63 (2013)
4. V.L. Deringer, W. Zhang, M. Lumeij, S. Maintz, M. Wuttig, R. Mazzarello, R. Dronskowski, Angew. Chem. Int. Ed. **53**, 10817–10820 (2014)
5. Y. Shirota, J. Mater. Chem. **10**, 1–25 (2000)
6. Y. Shirota, J. Mater. Chem. **15**, 75–93 (2005)
7. A. Mishra, P. Bäuerle, Angew. Chem. Int. Ed. **51**, 2020–2067 (2012)
8. J.C.S. Costa, L.J. Santos, Phys. Chem. C **117**, 10919–10928 (2013)
9. H. Nakano, T. Tanino, T. Takahashi, H. Ando, Y. Shirota, J. Mater. Chem. **18**, 242–246 (2008)
10. H. Nakano, S. Seki, H. Kageyama, Phys. Chem. Chem. Phys. **12**, 7772–7774 (2010)
11. Z. Li, Q. Dong, Y. Li, B. Xu, M. Deng, J. Pei, J. Zhang, F. Chen, S. Wen, Y. Gao, W. Tian, J. Mater. Chem. **21**, 2159–2168 (2011)
12. M. Grucela-Zajac, K. Bijak, S. Kula, M. Filapek, M. Wiacek, H. Janeczek, L. Skorka, J. Gasiorowski, K. Hingerl, N.S. Sariciftci, N. Nosidlak, G. Lewinska, J. Sanetra, E. Schab-Balcerzak, J. Phys. Chem. C **118**, 13070–13086 (2014)
13. J.-C.G. Bünzli, Chem. Rev. **110**, 2729–2755 (2010)
14. S.V. Eliseeva, J.-C.G. Bünzli, Chem. Soc. Rev. **39**, 189–227 (2010)
15. P. Nockemann, E. Beurer, K. Driesen, R. Van Deun, K. Van Hecke, L. Van Meervelt, K. Binnemans, Chem. Commun. **34**, 4354–4356 (2005)
16. S. Petoud, G. Muller, E.G. Moore, J.D. Xu, J. Sokolnicki, J.P. Riehl, U.N. Le, S.M. Cohen, K.N. Raymond, J. Am. Chem. Soc. **129**, 77–83 (2007)
17. S.J. Butler, D. Parker, Chem. Soc. Rev. **42**, 1652–1666 (2013)
18. A.P. Bassett, S.W. Magennis, P.B. Glover, D.J. Lewis, N. Spencer, S. Parsons, R.M. Williams, L. De Cola, Z. Pikramenou, J. Am. Chem. Soc. **126**, 9413–9424 (2004)

References

19. L. Armelao, S. Quici, F. Barigelletti, G. Accorsi, G. Bottaro, M. Cavazzini, E. Tondello, Coord. Chem. Rev. **254**, 487–505 (2010)
20. M.R. Robinson, M.B. O'Regan, G.C. Bazan, Chem. Commun. **254**, 1645–1646 (2000)
21. Y. Hasegawa, K. Murakoshi, Y. Wada, S. Yanagida, J.H. Kim, N. Nakashima, T. Yamanaka, Chem. Phys. Lett. **248**, 8–12 (1996)
22. Y. Hasegawa, K. Sogabe, Y. Wada, T. Kitamura, N. Nakashima, S. Yanagida, Chem. Lett. **1**, 35–36 (1999)
23. H.F. Brito, O.M.L. Malta, M.C.F.C. Felinto, E.E.S. Teotonio, Chem. Met. Enolates, chapter 3, 131 (2009)
24. N.B.D. Lima, S.M.C. Gonçalves, S.A. Junior, A.M. Simas, Sci. Rep. **3**, 1–8 (2013)
25. S.M. Borisov, O.S. Wolfbeis, Anal. Chem. **78**, 5094–5101 (2006)
26. G.E. Khalil, K. Lau, G.D. Phelan, B. Carlson, M. Gouterman, J.B. Callis, L.R. Dalton, Rev. Sci. Instrum. **75**, 192–206 (2004)
27. S.V. Eliseeva, D.N. Pleshkov, K.A. Lyssenko, L.S. Lepnev, J.-C.G. Bünzli, N.P. Kuzmina, Inorg. Chem. **49**, 9300–9311 (2010)
28. K. Nakamura, Y. Hasegawa, H. Kawai, N. Yasuda, Y. Wada, S. Yanagida, J. Alloys. Compd. **408**, 771–775 (2006)
29. K. Miyata, T. Ohba, A. Kobayashi, M. Kato, T. Nakanishi, K. Fushimi, Y. Hasegawa, ChemPlusChem **77**, 277–280 (2012)
30. K. Miyata, Y. Konno, T. Nakanishi, A. Kobayashi, M. Kato, K. Fushimi, Y. Hasegawa, Angew. Chem. Int. Ed. **52**, 6413–6416 (2013)
31. I.V. Alabugin, S.V. Kovalenko, J. Am. Chem. Soc. **124**, 9052–9053 (2002)
32. P. Suresh, S. Srimurugan, B. Babu, H.N. Pati, Acta Chim. Slov. **55**, 453–457 (2008)
33. J. Tirado-Rives, W.L. Jorgensen, J. Chem. Theory Comput. **4**, 297–306 (2008)
34. M. Dolg, H. Stoll, A. Savin, H. Preuss, Theor. Chim. Acta **75**, 173–194 (1989)
35. L. Maron, O. Einstein, J. Phys. Chem. A **104**, 7140–7143 (2000)
36. M.J. Frisch et al. *Gaussian 09*, Revision D.01. (Gaussian, Inc., Wallingford CT, 2009)
37. J. Ueda, S. Tanabe, J. Am. Ceram. Soc. **93**, 3084–3087 (2010)
38. E. Ueta, H. Nakano, Y. Shirota, Chem. Lett. **23**, 2397–2400 (1994)
39. M.D. Ediger, C.A. Angell, S.R. Nagel, J. Phys. Chem. **100**, 13200–13212 (1996)
40. M. Todoki, Sen'i Gakkaishi **65**, 385–393 (2009)
41. K. Nakamura, Y. Hasegawa, Y. Wada, S. Yanagida, Chem. Phys. Lett. **398**, 500–504 (2004)

Chapter 5
Amorphous Formability and Temperature-Sensitive Luminescence of Lanthanide Coordination Glasses

Abstract Glass-transition properties and temperature-sensitive luminescence of lanthanide (Ln(III)) coordination compounds are demonstrated. The amorphous formability was systematically provided by introducing bent-angled phosphine oxide ligands based on thienyl, naphthyl, and phenyl cores with ethynyl groups. Glass transition points were clearly identified for all Ln(III) coordination compounds from 65 to 87 °C. The Tb(III)/Eu(III) mixed coordination glass also exhibited temperature-dependent emission profiles from green, yellow, orange, to red in the range of 100–400 K.

Keywords Luminescence · Glass transition · Amorphous · Energy transfer

5.1 Introduction

Amorphous molecular materials or molecular glasses composed of low molecular-weight organic compounds are promising candidates for optical devices such as displays, lighting devices and LEDs because of their isotropy, transparency, and high processability as mentioned in Chap. 4 [1–3]. Nakano and Shirota provided a series of C_3 starburst organic compounds with glass transition and electrical conductivity [3, 4]. Bhowmik and co-workers demonstrated fluorescence and amorphous properties of molecular materials based on quinoline amines [5]. Molecular glasses are expected to be used as electro- and photo-active materials.

Luminescent Ln(III) coordination compounds are attractive candidates for optical applications such as displays and sensors [6–9]. They are usually dispersed in polymer matrices such as polymethyl methacrylate for transparent luminescent materials [10, 11]. In Chap. 4, the author reported the syntheses of luminescent and amorphous Ln(III) coordination compounds with Eu(III) ions, light-harvesting hexafluoroacetylacetonate (hfa) ligands, and 120°-angled bridging ligands with ethynyl groups (Fig. 5.1, *m*-dpeb) [12]. Quantum calculations and mass spectrometry revealed the formation of a *pseudo*-C_3 symmetrical trimer structure (Fig. 5.1, [Eu(hfa)$_3$(*m*-dpeb)]$_3$). The glass formability should be caused by the thermodynamically non-

Fig. 5.1 **a** The molecular design of Ln(III) coordination glasses and **b** chemical structures of organic ligands

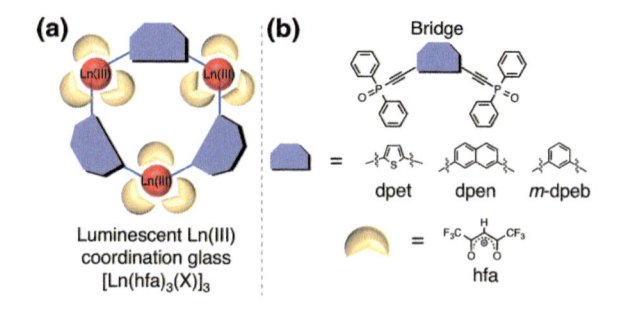

equilibrium states that arise from multiple quasi-stable structures, suppressing easy crystal packing to form an amorphous solid.

In this chapter, the author focuses on systematic construction of Ln(III) coordination compounds with amorphous-forming ability by simply altering the aryl cores of organic bridging ligands to broaden the range of molecular designs. Ln(III)-mixed coordination glasses would also provide transparent and temperature-dependent luminescent materials. Hasegawa and co-workers reported that Tb(III)/Eu(III) mixed rigid coordination polymers "chameleon luminophores" show brilliant green, yellow, orange, and red luminescence depending on temperature [13]. The emission colors and temperature sensitivity were also tuned by Tb(III)/Eu(III) mixture ratios in Chap. 3 [14]. These compounds are expected to be temperature/pressure sensitive paints (TSPs/PSPs) in the field of fluid dynamics and aeronautical engineering to visualize fluid phenomena such as aerodynamic heating [15–18]. The Ln(III)-mixed coordination glasses would provide a dramatically improved coating properties on the surface of materials for high-concentration TSPs/PSPs without polymer matrices.

In this chapter, the construction of Ln(III) coordination glasses and control of the physical properties are demonstrated. Three bidentate phosphine oxide ligands based on thienyl, naphthyl, and phenyl cores with ethynyl groups (2,5-bis(diphenylphosphorylethynyl)thiophene: dpet, 2,7-bis(diphenylphosphorylethynyl)naphthalene: dpen, and 1,3-bis(diphenylphosphorylethynyl)benzene: *m*-dpeb, Fig. 5.1b) were prepared. The corresponding Eu(III) coordination glasses were successfully synthesized, and their thermal properties were characterized by DSC and TG-DTA measurements. The emission properties were evaluated on the basis of photophysical parameters such as emission quantum yields and lifetimes. The temperature-sensitive luminescence of a Tb(III)/Eu(III) mixed coordination glass is also demonstrated.

5.2 Experimental Section

5.2.1 General

Europium acetate n-hydrate (99.9%), terbium(III) acetate tetrahydrate (99.9%), n-BuLi (in n-hexane, 1.6 M), and hydrogen peroxide were purchased from Kanto Chemical Co., Inc. 2,5-Dibromothiophene, 3,4-ethylenedioxythiophene, and chlorodiphenylphosphine were obtained from Tokyo Chemical Industry Co., Ltd. All other chemicals and solvents were reagent grade and were used without further purification.

5.2.2 Apparatus

All chemicals are reagent grade and used without further purification. ^{1}H NMR (400 MHz) spectra were recorded on a JEOL ECS400. Chemical shifts were reported in δ ppm, referenced to an internal tetramethylsilane standard for ^{1}H NMR spectroscopy. Infrared spectra were recorded on a JASCO FTIR-420 spectrometer using KBr pellets. Elemental analyses were performed by an Exeter Analytical CE440. Mass spectrometry was performed by a Thermo Scientific Exactive (ESI-MS) and a JEOL JMS-700TZ (FAB-MS). Differential thermal analysis was performed on a Shimadzu DSC-60 Plus under a nitrogen atmosphere at a heating/cooling rate of 5 °C min^{-1}. Thermogravimetric analyses were conducted on a Shimadzu DTG-60 under a nitrogen atmosphere at a heating/cooling rate of 5 °C min^{-1}.

5.2.3 Syntheses

The organic bridging ligands and corresponding coordination glasses were synthesized following Scheme 5.1.

Preparation of 2,5-bis(trimethylsilylethynyl)thiophene:

Trimethylsilylacetylene (9.5 mL, 69 mmol) was added in one portion to a solution of 2,5-dibromothiophene (3.0 mL, 27 mmol), PdCl$_2$(PPh$_3$)$_2$ (0.98 g), CuI (0.34 g), PPh$_3$ (0.94 g) in diisopropylamine/dry THF (80 mL/80 mL) under argon atmosphere and stirred at room temperature for 20 min. The mixture was stirred at 50 °C for 9 h, and then allowed to cool to room temperature. The ammonium salt was removed by filtration and the mixture was extracted with hexane and concentrated. The obtained crude oil was purified by column chromatography on SiO$_2$ using hexane as an eluent to afford 2,5-bis(trimethylsilylethynyl)thiophene (Yield: 6.5 g, 89%).

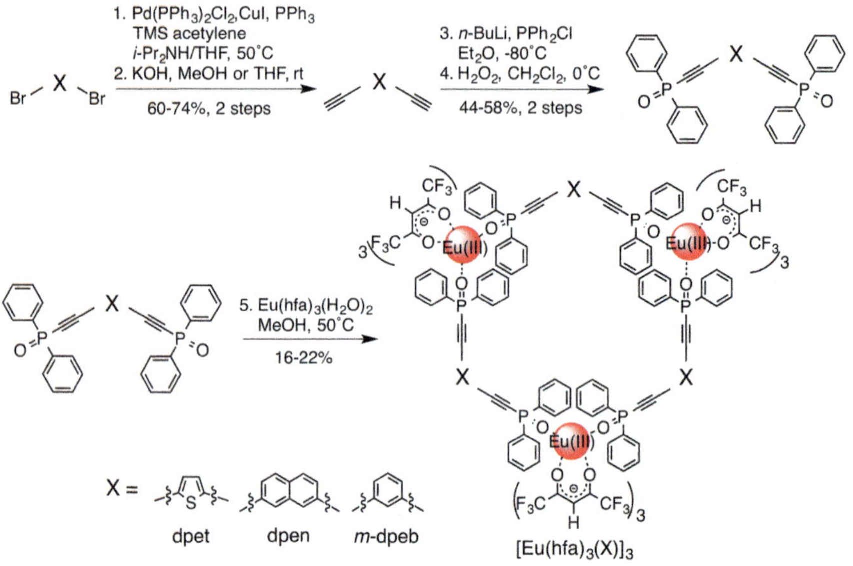

Scheme 5.1 Synthetic schemes of phosphine oxide ligands with ethynyl groups and Eu(III) coordination glasses

Preparation of 2,5-diethynylthiophene:

The resulting 2,5-bis(trimethylsilylethynyl)thiophene (6.5 g, 24 mmol) was dissolved in methanol (100 mL) and 1 M KOH aqueous solution (60 mL, 60 mmol) was added. The mixture was stirred for 3 h. The product was extracted with hexane, concentrated, and purified by column chromatography on SiO_2 using hexane as an eluent to afford 2,5-diethynylthiopehene (Yield: 2.1 g, 67%).

Preparation of 2,5-bis(diphenylphosphorylethynyl)thiophene (dpet):

The 2,5-diethynylthiophene (2.1 g, 16 mmol) was dissolved in dry Et_2O (80 mL) under argon atmosphere. A solution of 1.6 M n-BuLi (23 mL, 35 mmol) was added dropwise to the solution at −80 °C. The mixture was allowed to stir for 3 h at −10 °C, after which a PPh_2Cl (6.2 mL, 34 mmol) was added dropwise at −80 °C. The mixture was gradually brought to room temperature, and stirred overnight. The product was extracted with dichloromethane, concentrated, and re-dissolved in dichloromethane (60 mL), followed by addition of a 30% H_2O_2 aqueous solution (11 mL). The reaction mixture was stirred for 3 h at 0 °C. The product was extracted with dichloromethane and purified by column chromatography on SiO_2 using ethyl acetate and hexane as mixed eluent (ethyl acetate:hexane = 2:1). The residue was concentrated and recrystallized in ethyl acetate. The obtained powder was washed with ethyl acetate.

Yield: 3.6 g (26%). ^{1}H NMR (400 MHz, $CDCl_3$, 25 °C) δ 7.83–7.89 (m, 8H, –CH), δ 7.49–7.62 (m, 12H, –CH), δ 7.35 (s, 2H, –CH) ppm. ESI-Mass (*m/z*): calcd for $C_{32}H_{23}O_2P_2S$ [M+H]$^+$, 533.09; found, 533.09. Anal. Calcd for $C_{32}H_{22}O_2P_2S$: C, 72.17; H, 4.16%. Found: C, 72.31; H, 4.37%.

Preparation of 2,7-bis(trimethylsilylethynyl)naphthalene:

Trimethylsilylacetylene (5.5 mL, 40 mmol) was added in one portion to a solution of 2,7-dibromonaphthalene (5.3 g, 19 mmol), PdCl$_2$(PPh$_3$)$_2$ (0.55 g), CuI (0.19 g), PPh$_3$ (0.53 g) in diisopropylamine/dry THF (40 mL/100 mL) under argon atmosphere and stirred at room temperature for 20 min. The mixture was stirred at 50 °C overnight, and then allowed to cool to room temperature. The ammonium salt was removed by filtration and the mixture was extracted with ethyl acetate and concentrated. The obtained crude oil was purified by column chromatography on SiO$_2$ using hexane and ethyl acetate as mixed eluent (hexane:ethyl acetate = 5:1) to afford 2,7-bis(trimethylsilylethynyl)naphthalene (Yield: 5.1 g, 84%).

Preparation of 2,7-diethynylnaphthalene:

2,7-Bis(trimethylsilylethynyl)naphthalene (5.1 g, 16 mmol) was dissolved in THF (100 mL) and 1 M KOH aqueous solution (45 mL, 45 mmol) was added. The mixture was stirred overnight. The product was extracted with hexane and concentrated to afford 2,7-diethynylnaphthalene (Yield: 2.5 g, 88%).

Preparation of 2,7-bis(diphenylphosphorylethynyl)naphthalene:

The 2,7-diethynylnaphthalene (2.5 g, 14 mmol) was dissolved in dry Et$_2$O (80 mL) under argon atmosphere. A solution of 1.6 M n-BuLi (20 mL, 31 mmol) was added dropwise at −80 °C. The mixture was allowed to stir for 3 h at −10 °C, after which a PPh$_2$Cl (5.6 mL, 31 mmol) was added dropwise at −80 °C. The mixture was gradually brought to room temperature and stirred overnight. The product was extracted with dichloromethane, concentrated, and re-dissolved in dichloromethane (60 mL), followed by addition of a 30% H$_2$O$_2$ aqueous solution (10 mL). The reaction mixture was stirred for 3 h at 0 °C. The product was extracted with dichloromethane and recrystallized. The obtained crude powder was washed with ethyl acetate.

Yield: 3.6 g (34%). ^{1}H NMR (400 MHz, CDCl$_3$, 25 °C) δ 8.12 (s, 2H, –CH), δ 7.89–7.95 (m, 8H, –CH), δ 7.82–7.84 (d, 2H, –CH), δ 7.64–7.66 (dd, 2H, –CH), δ 7.48–7.59 (m, 12H, –CH) ppm. ESI-Mass (m/z): calcd for $C_{38}H_{27}O_2P_2$ [M+H]$^+$, 577.15; found, 577.15. Anal. Calcd for $C_{38}H_{26}O_2P_2$: C, 79.16; H, 4.55%. Found: C, 79.05; H, 4.70%.

Preparation of Eu(III) coordination glasses [Eu(hfa)$_3$(X)]$_3$ (X = dpet, dpen, m-dpeb):

The phosphine oxide ligand (0.8 mmoL) and Eu(hfa)$_3$(H$_2$O)$_2$ (0.8 mmoL) was dissolved in methanol (40 mL), respectively. The solutions were mixed and stirred for 3 h at 50 °C. The reaction mixture was concentrated, re-dissolved in methanol (5 mL), and then hexane (15 mL) was added. The organic solvents were removed by decompression to form amorphous solids.

[**Eu(hfa)₃(dpet)**]₃. Yield: 0.62 g (20%). IR (KBr) 1656 (st, C=O), 1144 (st, P=O), 1100–1253 (st, C–O–C and st, C–F) cm⁻¹. FAB-Mass (m/z): [M-hfa]⁺ calcd for $C_{136}H_{74}Eu_3F_{48}O_{22}P_6S_3$, 3710.9; found, 3709.9. Anal. Calcd for $C_{141}H_{75}Eu_3F_{54}O_{24}P_6S_3$: C, 43.24; H, 1.93%. Found: C, 44.02; H, 2.24%.

[**Eu(hfa)₃(dpen)**]₃. Yield: 0.52 g (16%). IR (KBr) 1656 (st, C=O), 1144 (st, P=O), 1100–1256 (st, C–O–C and st, C–F) cm⁻¹. FAB-Mass (m/z): [M-hfa]⁺ calcd for $C_{154}H_{86}Eu_3F_{48}O_{22}P_6$, 3843.1; found, 3842.0. Anal. Calcd for $C_{159}H_{87}Eu_3F_{54}O_{24}P_6$: C, 47.17; H, 2.17%. Found: C, 46.85; H, 2.39%.

5.2.4 Optical Measurements

UV-Vis absorption spectra were recorded on a JASCO V-670 spectrometer. Emission and excitation spectra were recorded on a HORIBA Fluorolog-3 spectrofluorometer and corrected for the response of the detector system. The temperature-dependent emission spectra were measured with a nitrogen bath cryostat (Oxford Instruments, Optistat DN) and a temperature controller (Oxford, Instruments, ITC 502S). Emission lifetimes (τ_{obs}) were measured using the third harmonics (355 nm) of a Q-switched Nd:YAG laser (Spectra Physics, INDI-50, fwhm = 5 ns, λ = 1064 nm) and a photomultiplier (Hamamatsu photonics, R5108, response time ≤ 1.1 ns). The Nd:YAG laser response was monitored with a digital oscilloscope (Sony Tektronix, TDS3052, 500 MHz) synchronized to the single-pulse excitation. Emission lifetimes were determined from the slope of logarithmic plots of the decay profiles. The emission quantum yields excited at 380 nm (Φ_{tot}) were estimated using JASCO F-6300-H spectrometer attached with a JASCO ILF-533 integrating sphere unit ($\varphi = 100$ mm). The wavelength dependence of the detector response and the beam intensity of Xe light source for each spectrum were calibrated using a standard light source.

5.3 Results and Discussion

5.3.1 Structural Characterizations

Fragments of trimer structures [Eu₃(hfa)₈(X)₃]⁺ (X = dpet, dpen, m-dpeb) in the mass spectra were confirmed for all compounds. The peaks for larger m/z values also indicate the presence of a wide range of other cyclic or linear oligomers. Considering the results of mass spectrometry and the DFT optimized structure as reported for [Eu(hfa)₃(m-dpeb)]₃ [12], the author assumes that all of these compounds form trimer structures with a *pseudo-C₃* axis in an amorphous solid state. The glassy states of these compounds were also demonstrated by halo signals of XRD patterns (Fig. 5.2).

Fig. 5.2 X-ray diffraction patterns of Eu(III) coordination glasses, **a** [Eu(hfa)$_3$(dpet)]$_3$ and **b** [Eu(hfa)$_3$(dpen)]$_3$

Fig. 5.3 DSC thermograms of Eu(III) coordination glasses, **a** [Eu(hfa)$_3$(dpet)]$_3$, **b** [Eu(hfa)$_3$(dpen)]$_3$, and **c** [Eu(hfa)$_3$(*m*-dpeb)]$_3$ (dashed line: cooling process, solid line: heating process), the chemical structures on right shows the bridging ligands

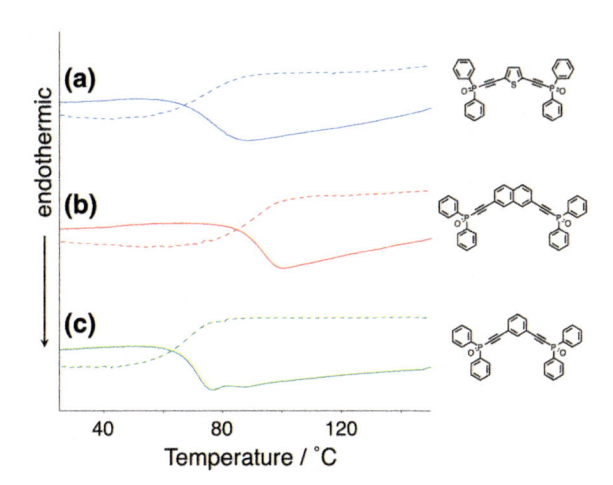

5.3.2 Thermal Properties

DSC measurements were performed to determine the glass transition temperatures. All of the compounds clearly showed glass-transition temperatures, which were characteristic of amorphous molecular materials (Fig. 5.3; Table 5.1). These compounds formed transparent viscous fluids above T_g, and the intermolecular rearrangements of broken C_3-symmetrical units were probably frozen below T_g in a cooling process. The author considers the highest T_g of [Eu(hfa)$_3$(dpen)]$_3$ ($T_g = 87$ °C) to be responsible for the largest molecular weight, planarity, and availability of aromatic surfaces for intermolecular interactions of the bridging ligand, dpen. The ethynyl groups in bridging ligands also play an important role in suppressing tight-binding interaction and crystallization of assembled Eu(III) coordination compounds by enlarging the cell volume without extending π-aromatic surface for intermolecular interactions, which also lead to relatively low thermal decomposition points (Fig. 5.4; Table 5.1).

The reported thiophene-based Eu(III) coordination polymer without ethynyl groups, [Eu(hfa)$_3$(dpt)]$_n$ (dpt: 2,5-bis(diphenylphosphoryl)thiophene) [19], exhib-

Table 5.1 Thermal properties of an Eu(III) coordination polymer [Eu(hfa)$_3$(dpt)]$_n$ and coordination glasses [Eu(hfa)$_3$(X)]$_3$

Compounds	M_w of bridging ligand (g mol^{-1})	Glass transition temperature (°C)	Decomposition temperature (°C)
[Eu(hfa)$_3$(dpt)]$_n^a$	484.08	–	322
[Eu(hfa)$_3$(dpet)]$_3$	532.08	73	242
[Eu(hfa)$_3$(dpen)]$_3$	576.14	87	260
[Eu(hfa)$_3$(m-dpeb)]$_3^b$	526.13	65	249

[a]Reference [19], [b]Reference [12]

Fig. 5.4 TG curves of Eu(III) coordination glasses, **a** Eu(hfa)$_3$(dpet)]$_3$, **b** [Eu(hfa)$_3$(dpen)]$_3$, and **c** [Eu(hfa)$_3$(m-dpeb)]$_3$ (5 °C min^{-1}, N$_2$ atmosphere), the chemical structures on right show the bridging ligands

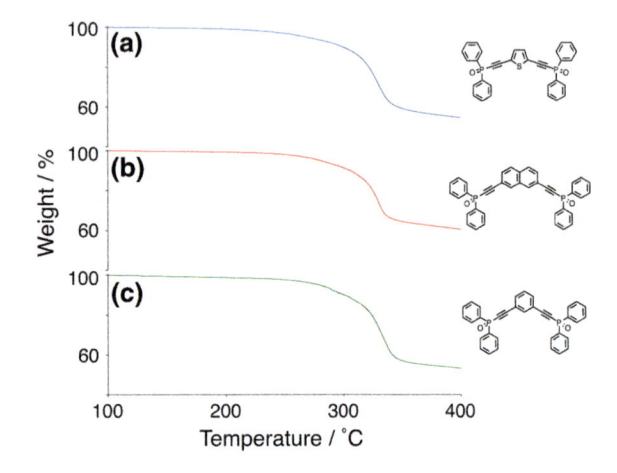

Fig. 5.5 A DSC curve of [Eu(hfa)$_3$(dpt)]$_n$ (dashed line: cooling process, solid line: 2nd heating process), the chemical structure on right shows the bridging ligands

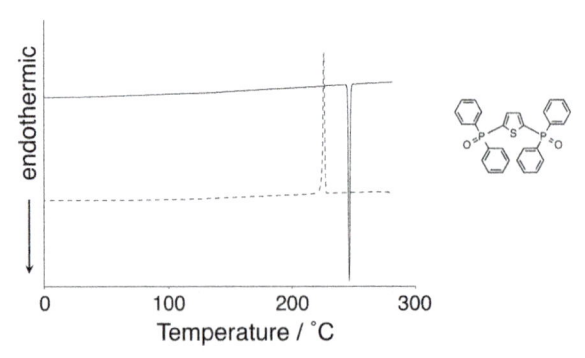

ited no glass-transition point (Fig. 5.5). Thus, glass-forming ability can be systematically introduced by bent-angled bridging ligands with ethynyl groups that make more irregularly shaped molecules, which is more difficult to crystallize and more likely to form amorphous state.

5.3.3 Photophysical Properties

Diffuse-reflectance absorption spectra and emission spectra were measured to evaluate the photophysical properties of Eu(III) coordination glasses (Fig. 5.6). The absorption bands at around 350 nm were assigned to the π–π* transitions of antenna ligands. Absorption shoulder bands of the bridging ligands were also observed at around 500 nm for [Eu(hfa)$_3$(dpet)]$_3$ and [Eu(hfa)$_3$(dpen)]$_3$. Sharp and small absorption bands at 465 nm were due to the $^7F_0 \rightarrow {}^5D_2$ transitions of Eu(III) ions. The emission bands at 578, 591, 613, 650, and 698 nm are attributed to the 4f–4f transitions of $^5D_0 \rightarrow {}^7F_J$ ($J = 0, 1, 2, 3, 4$). The emission lifetimes (τ_{obs}) were measured and analyzed as single-exponential decay with a millisecond-scale lifetime for each Eu(III) coordination glass. Their intrinsic emission quantum yields (Φ_{ff}), radiative (k_r), and non-radiative constants (k_{nr}) were estimated using Eqs. (3–5) as shown in Chap. 2. The estimated photophysical parameters are summarized in Table 5.2.

The Φ_{ff} values of [Eu(hfa)$_3$(dpen)]$_3$ and [Eu(hfa)$_3$(m-dpeb)]$_3$ were estimated to be 46 and 72%, respectively. Unfortunately, [Eu(hfa)$_3$(dpet)]$_3$ exhibits weak red luminescence along with the luminescence from bridging ligands in the visible region, and thus the photophysical parameters were not available. The k_r value

Fig. 5.6 Diffuse-reflectance absorption spectra (left) and emission spectra excited at 380 nm (right) of Eu(III) coordination glasses, **a** [Eu(hfa)$_3$(dpet)]$_3$, **b** [Eu(hfa)$_3$(dpen)]$_3$, and **c** [Eu(hfa)$_3$(m-dpeb)]$_3$

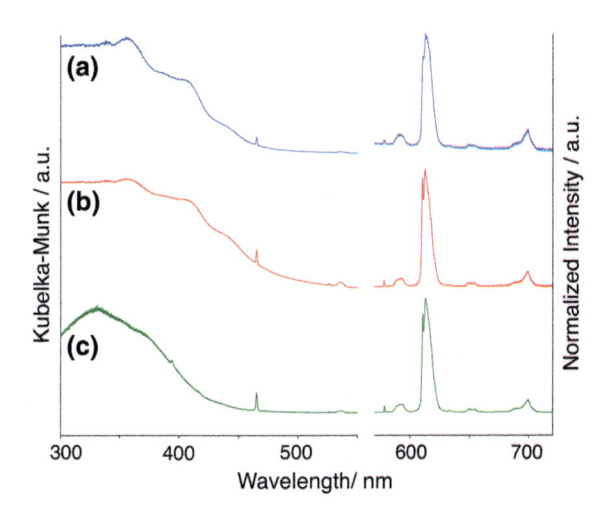

Table 5.2 Photophysical parameters of Eu(III) coordination glasses [Eu(hfa)$_3$(X)]$_3$ in solid state

X	Φ_{tot}^a (%)	Φ_{ff}^b (%)	η_{sens}^c (%)	τ_{obs}^c (ms)	k_r^b (s^{-1})	k_{nr}^b (s^{-1})
dpet[d]	n.a.	n.a.	n.a.	0.06	n.a.	n.a.
dpen	15	46	33	0.63	7.2×10^2	8.6×10^2
m-dpeb[e]	44	72	61	1.0	7.2×10^2	2.8×10^2

[a]λ_{ex} = 380 nm, [b]Equations (3–5) in Chap. 2, [c]λ_{ex} = 355 nm, [d]Photophysical parameters were not available because of the weak luminescence, [e]Reference [12]

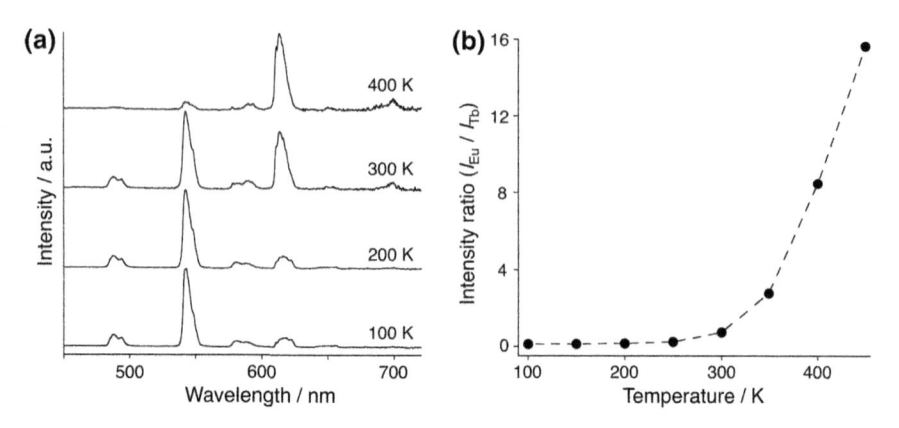

Fig. 5.7 Temperature-dependent, **a** emission spectra and **b** emission intensity ratio of [Tb,Eu(hfa)$_3$(*m*-dpeb)]$_3$ (Tb/Eu = 50) in amorphous state (λ_{ex} = 380 nm)

of [Eu(hfa)$_3$(dpen)]$_3$ was the same as that of [Eu(hfa)$_3$(*m*-dpeb)]$_3$, indicating that the coordination geometries around Eu(III) ions in the trimer structures were similar. The k_{nr} value of [Eu(hfa)$_3$(dpen)]$_3$ was much larger than that of [Eu(hfa)$_3$(*m*-dpeb)]$_3$. The low-lying absorption edges of [Eu(hfa)$_3$(dpen)]$_3$ and [Eu(hfa)$_3$(dpet)]$_3$ at around 500 nm indicated the low excited triplet (T$_1$) states. The relatively lower T$_1$ states of dpen and dpet would promote the quenching of Eu(III)-centered emission through ligand-centered emission [23]. Thus, the electronic structures of organic cores in the bridging ligands affect the Eu(III)-centered emission properties of coordination glasses.

A Tb(III)/Eu(III)-mixed coordination glass, [Tb,Eu(hfa)$_3$(*m*-dpeb)]$_3$ (Tb/Eu = 50), was also prepared to introduce a temperature-responsive emission property in a glassy state. The emission spectra were recorded in the range of 100–400 K, and corresponding green to red photoluminescence in glassy state was observed (Fig. 5.7a). According to the emission intensity ratio (Fig. 5.7b), the Tb(III)/Eu(III)-mixed coordination glass was found to show as high temperature sensitivity (0.92% K^{-1}) as that of the reported chameleon luminophore, [Tb,Eu(hfa)$_3$(dpbp)]$_3$ (Tb/Eu = 99, 0.83% K^{-1}) [13]. These results indicate that energy transfer between Tb(III) and Eu(III) ions can effectively occur in the amorphous state.

5.4 Conclusions

A series of Ln(III) coordination glasses was successfully prepared, and temperature-sensitive luminescence in a glassy state was reported for the first time. The glass transition properties were dominated by molecular weight, geometry of the bridging ligands, and availability of aromatic surface for π-stacking. The trimer structures of

5.4 Conclusions

Ln(III) coordination compounds were found to be suitable for thermodynamically non-equilibrium states, which makes them more difficult to crystallize and more likely to form an amorphous solid. The Tb(III)/Eu(III) mixed coordination glass also showed green, yellow, orange and red photoluminescence depending on temperature.

The idea of coordination glass enables the reported emissive Ln(III) coordination compounds to achieve dramatically improved painting properties on the surface of materials. These molecular designs also have the potential for providing thermo-stable Ln(III) coordination glasses for industrial applications. There is a global demand for these compounds as highly processable emissive dyes for fluidics, aeronautics, and chemical engineering.

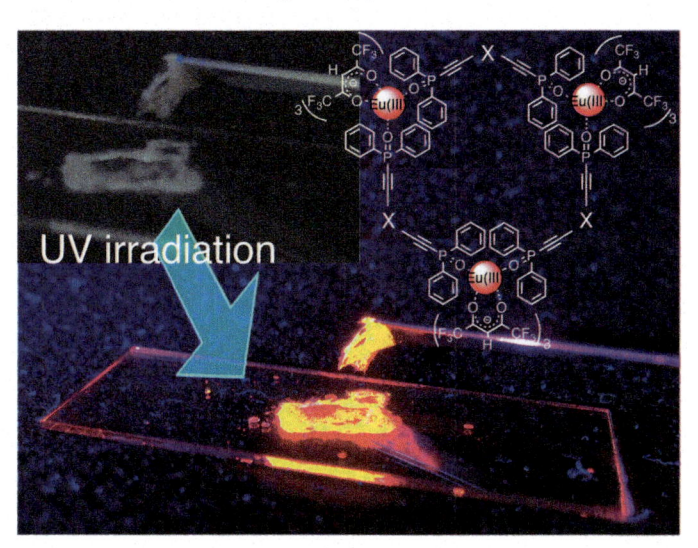

References

1. A. Mishra, P. Bauerle, Angew. Chem. Int. Ed. **51**, 2020–2067 (2012)
2. J.C.S. Costa, L. Santos, J. Phys. Chem. C **117**, 10919–10928 (2013)
3. Y. Shirota, J. Mater. Chem. **10**, 1–25 (2000)
4. H. Nakano, S. Seki, H. Kageyama, Phys. Chem. Chem. Phys. **12**, 7772–7774 (2010)
5. A.K. Nedeltchev, H. Han, P.K. Bhowmik, Tetrahedron **66**, 9319–9326 (2010)
6. J. Kido, Y. Okamoto, Chem. Rev. **102**, 2357–2368 (2002)
7. J.-C.G. Bünzli, Chem. Rev. **110**, 27292755 (2010)
8. A. de Bettencourt-Dias, *Dalton Trans*, 2229–2241 (2007)
9. S.J. Butler, D. Parker, Chem. Soc. Rev. **42**, 1652–1666 (2013)
10. O. Moudam, B.C. Rowan, M. Alamiry, P. Richardson, B.S. Richards, A.C. Jones, N. Robertson, Chem. Commun, 6649–6651 (2009)
11. G. Zucchi, V. Murugesan, D. Tondelier, D. Aldakov, T. Jeon, F. Yang, P. Thuery, M. Ephritikhine, B. Geffroy, Inorg. Chem. **50**, 4851–4856 (2011)
12. Y. Hirai, T. Nakanishi, Y. Kitagawa, K. Fushimi, T. Seki, H. Ito, H. Fueno, K. Tanaka, T. Satoh, Y. Hasegawa, Inorg. Chem. **54**, 4364–4370 (2015)

13. K. Miyata, Y. Konno, T. Nakanishi, A. Kobayashi, M. Kato, K. Fushimi, Y. Hasegawa, Angew. Chem. Int. Ed. **52**, 6413–6416 (2013)
14. Y. Hirai, T. Nakanishi, K. Miyata, K. Fushimi, Y. Hasegawa, Mater. Lett. **130**, 91–93 (2014)
15. S.M. Borisov, O.S. Wolfbeis, Anal. Chem. **78**, 5094–5101 (2006)
16. M. Schaferling, Angew. Chem. Int. Ed. **51**, 3532–3554 (2012)
17. M.I.J. Stich, S. Nagl, O.S. Wolfbeis, U. Henne, M. Schaeferling, Adv. Funct. Mater. **18**, 1399–1406 (2008)
18. K. Nakakita, M. Kurita, K. Mitsuo, S. Watanabe, Meas. Sci. Technol. **17**, 359–366 (2008)
19. Y. Hirai, T. Nakanishi, Y. Kitagawa, K. Fushimi, T. Seki, H. Ito, Y. Hasegawa, Angew. Chem. Int. Ed. **55**, 12059–12062 (2016)
20. M.H.V. Werts, R.T.F. Jukes, J.W. Verhoeven, Phys. Chem. Chem. Phys. **4**, 1542–1548 (2002)
21. A. Aebischer, F. Gumy, J.-C.G. Bünzli, Phys. Chem. Chem. Phys. **11**, 1346–1353 (2009)
22. R. Pavithran, N.S.S. Kumar, S. Biju, M.L.P. Reddy, S.A. Junior, R.O. Freire, Inorg. Chem. **45**, 2184–2192 (2006)
23. M. Hatanaka, Y. Hirai, Y. Kitagawa, T. Nakanishi, Y. Hasegawa, K. Morokuma, Chem. Sci. **8**, 423–429 (2017)

Chapter 6
Triboluminescence of Lanthanide Coordination Polymers

Abstract Triboluminescence (TL) and photoluminescence (PL) of novel lanthanide (Ln(III)) coordination polymers [Ln(hfa)$_3$(dpf)]$_n$ (dpf: 2,5-bis(diphenylphosphoryl)furan, Ln=Tb, Gd, Eu) are reported. The coordination polymers exhibited bright TL due to the face-to-face arrangement of substituents between single polymer chains. The observation of TL in a series of Eu(III) coordination polymers and a Gd(III) compound indicated that both hfa ligands and Ln(III) ions were excited under grinding. Significant PL/TL spectral differences in [Tb,Eu(hfa)$_3$(dpf)]$_n$ due to distinct excitation processes upon grinding and UV irradiation were observed for the first time.

Keywords Triboluminescence · Photoluminescence · Coordination polymer

6.1 Introduction

Luminescence upon grinding solid materials is called triboluminescence (TL). In 17th century, it was first noted by Francis Bacon that: "It is well known that all sugar, whether candied or plain, if it be hard, will sparkle when broken or scraped in the dark." TL phenomena, unlike well-known photoluminescence (PL), have advantages that they are generated without photo-irradiation. The fracture-induced luminescence properties of TL materials make them attractive for application as structural damage sensors and security marking techniques [1–4].

TL phenomena have been widely found in inorganic and organic compounds [5–9], and about half of all crystalline materials are predicted to show TL [10]. Wang demonstrated yellow-to-red TL of CaZnOS: Mn(II) depending on the concentration of Mn(II) ions [11]. Yamashita and co-workers reported blue TL of trifluoromethyl and pentafluoromethyl substituted imide derivatives under daylight [12, 13]. Coordination compounds including Cu(I), Pt(II), Mn(II), and Ln(III) ions have also been reported for their TL properties [14–18]. These TL-active compounds are known to show PL as well. TL and PL arise from distinct stimuli such as grinding and photo-irradiation; however, their spectra usually exhibit similar profiles [19, 20].

Relationships between TL and PL are still being discussed form a scientific point of view. In order to provide new insights into TL phenomena, the author focuses on Ln(III) complexes with efficient TL and PL properties because of their high spectral resolution arising from intraconfigurational 4f–4f transitions [21–25]. Reddy and co-workers recently demonstrated bright red TL and PL of [Eu(DPPF)$_3$(DDXPO)] (DPPF: 1-(4-(diphenylphosphino)phenyl)-4,4,5,5,5-pentafluoro-pentane-1,3-dione, DDXPO: 4,5-bis(diphenylphosphino)-9,9-dimethylxanthene oxide) [26]. They also reported slight decrease in TL intensity when dispersed in PMMA films, compared with the solid state.

The author here focused on the specific arrangement of substituents in solid-state Ln(III) complexes. Sweeting and Rheingold reported that the disorder of phenyl rings and cations in [Eu(dbm)$_4$]$^-$[Et$_3$NH]$^+$ (dbm: dibenzoylmethanato) provided a sufficient source of localized polarity to induce TL activity [27]. Chen reported that the disorder of thienyl and trifluoromethyl (CF$_3$) groups in Eu(III) complexes played an important role in providing charge separation by creating randomly distributed sites of electron affinities at the faces of fractures [17]. There have also been some reports on TL-active Ln(III) coordination polymers based on β-diketonato and aromatic bridging ligands [28, 29].

The author considers that strong TL Ln(III) coordination polymers can be constructed by introducing disordered arrangement of substituents between single polymer chains in the solid state. A novel furan-based bridging ligand (2,5-bis(diphenylphosphoryl)furan: dpf) is designed to provide the disordered face-to-face arrangement of CF$_3$ substituents in Ln(III)-hfa coordination polymers (Fig. 6.1a). The polar character and small aromatic ring of the furyl-bridging ligand are expected to prevent the polymer chains from forming highly ordered alternate packing structures (Fig. 6.1b). The reported coordination polymers with thiophene-based bridges, [Eu(hfa)$_3$(dpt)]$_n$ and [Eu(hfa)$_3$(dpedot)]$_n$ are also prepared for comparison of TL and PL properties [30]. The relationships between coordination structures and TL activities are evaluated using single crystal X-ray analyses and TL observations. A large spectral difference between TL and PL is confirmed in Tb(III)/Eu(III) mixed polymers using a charge coupled detector (CCD) system. The excitation processes in TL and PL are proposed on the basis of photo science.

6.2 Experimental Section

6.2.1 General

Europium acetate n-hydrate (99.9%), terbium(III) acetate tetrahydrate (99.9%), gadolinium(III) acetate tetrahydrate (99.9%), n-BuLi (in n-hexane, 1.6 M), and hydrogen peroxide were purchased from Kanto Chemical Co., Inc. 2,5-Dibromofuran, 3,4-ethylenedioxythiophene, 2,5-dibromothiophene, and chlorodiphenylphosphine were obtained from Tokyo Chemical Industry Co.,

6.2 Experimental Section

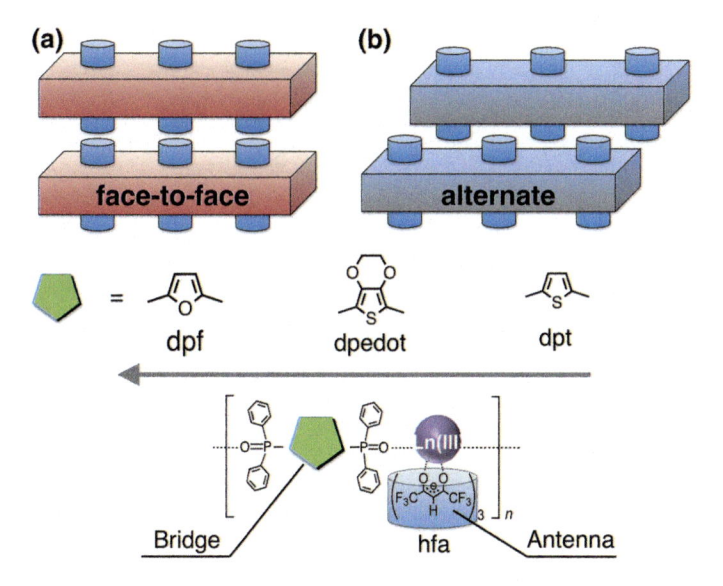

Fig. 6.1 Schematic representation of Ln(III) coordination polymers with, **a** face-to-face and **b** alternate intermolecular packing structures

Ltd. All other chemicals and solvents were reagent grade and were used without further purification.

6.2.2 Apparatus

^{1}H NMR (400 MHz) spectra were recorded on a JEOL ECS400. Chemical shifts were reported in δ ppm, referenced to an internal tetramethylsilane standard for ^{1}H NMR spectroscopy. Infrared spectra were recorded on a JASCO FTIR-420 spectrometer using KBr pellets. Elemental analyses were performed by an Exeter Analytical CE440. Mass spectrometry was performed by a Thermo Scientific Exactive (ESI-MS) and a JEOL JMS-700TZ (FAB-MS).

6.2.3 Syntheses

The organic bridging ligands and corresponding coordination polymers were synthesized following Scheme 6.1.

Preparation of 2,5-bis(diphenylphosphoryl)furan (dpf):

In a degassed 3-neck round-bottomed flask (300 mL vol.), 2,5-dibromofuran (2.6 mL, 24 mmol) was dissolved in dry THF (130 mL) under argon atmosphere,

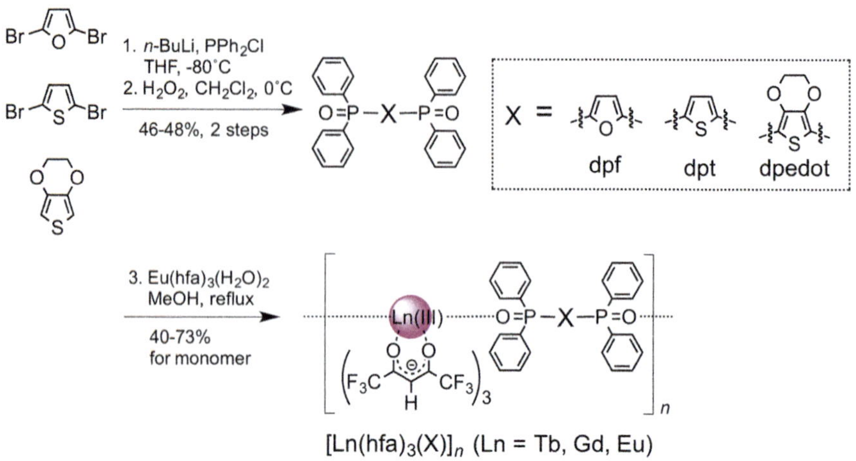

Scheme 6.1 Synthetic schemes of phosphine oxide ligands and Ln(III) coordination polymers

then stirred until a homogeneous solution was formed at room temperature. A solution of n-BuLi (40 mL, 62 mmol) was added dropwise to the solution at −80 °C. The addition was completed in ca. 15 min. The mixture was allowed to stir for 3 h, after which a PPh$_2$Cl (12 mL, 67 mmol) was added dropwise at −80 °C, then the solution became cloudy. The mixture was gradually brought to room temperature, and stirred for 14 h to form clouded yellow solution. The product was extracted with dichloromethane and dried over anhydrous MgSO$_4$. The solvent was concentrated and dissolved in dichloromethane (100 mL) in a flask. The solution was cooled to 0 °C and then 30% H$_2$O$_2$ aqueous solution (21 mL) was added to it. The reaction mixture was stirred for 3 h. The product was extracted with dichloromethane and the obtained crude powder was washed with ethyl acetate for several times to afford white powder.

Yield: 5.5 g (48%). ^{1}H NMR (400 MHz, CD$_3$OD, 25 °C) δ 7.61–7.67 (m, 12H, –CH), δ 7.51-7.55 (m, 8H, –CH), δ 7.27 (t, 2H, –CH) ppm. ESI-Mass (m/z): calcd for C$_{28}$H$_{23}$O$_3$P$_2$ [M+H]$^+$, 469.11; found, 469.11. Anal. Calcd for C$_{28}$H$_{22}$O$_3$P$_2$: C, 71.79; H, 4.73%. Found: C, 71.51; H, 4.67%.

Preparation of Ln(III) coordination polymers [Ln(hfa)$_3$(dpf)]$_n$ (Ln=Tb, Gd, Eu):

Phosphine oxide ligand, dpf (0.37 g, 0.80 mmol) and Ln(hfa)$_3$(H$_2$O)$_2$ (Ln=Tb, Gd, Eu, 0.80 mmol) were dissolved in methanol (40 mL), respectively. The solutions were mixed and refluxed for 3 h. The reaction mixture was concentrated and washed with chloroform. The solvent was evaporated and re-dissolved in 50 °C methanol for recrystallization. The obtained crystals were washed with −20 °C methanol.

[Tb(hfa)$_3$(dpf)]$_n$. Yield: 0.65 g (65% for monomer). IR (KBr): 1657 (st, C=O), 1145 (st, P=O) cm^{-1}. ESI-Mass (m/z): calcd for C$_{38}$H$_{24}$TbF$_{12}$O$_7$P$_2$ [M-hfa]$^+$: 1041.0, found, 1041.0. Anal. Calcd for $C_{43}H_{25}TbF_{18}O_9P_2$: C, 41.37; H, 2.02%. Found: C, 41.32; H, 1.88%.

[Gd(hfa)₃(dpf)]ₙ. Yield: 0.57 g (57% for monomer). IR (KBr): 1657 (st, C=O), 1145 (st, P=O) cm⁻¹. ESI-Mass (m/z): calcd for $C_{38}H_{24}GdF_{12}O_7P_2$ [M-hfa]⁺: 1040.0, found, 1040.0. Anal. Calcd for $C_{43}H_{25}GdF_{18}O_9P_2$: C, 41.42; H, 2.02%. Found: C, 41.30; H, 1.87%.

[Eu(hfa)₃(dpf)]ₙ. Yield: 0.73 g (73% for monomer). IR (KBr): 1657 (st, C=O), 1145 (st, P=O) cm⁻¹. ESI-Mass (m/z): calcd for $C_{38}H_{24}EuF_{12}O_7P_2$ [M-hfa]⁺: 1035.0, found, 1034.7. Anal. Calcd for $C_{43}H_{25}EuF_{18}O_9P_2$: C, 41.60; H, 2.03%. Found: C, 41.62; H, 2.32%.

Preparation of Tb(III)/Eu(III) mixed coordination polymers [Tb,Eu(hfa)₃(dpf)]ₙ (Tb/Eu = 1):

Tb(hfa)₃(H₂O)₂ (0.34 g, 0.40 mmol) and Eu(hfa)₃(H₂O)₂ (0.34 g, 0.40 mmol) were dissolved in methanol (30 mL). The Tb(III)/Eu(III)-mixed solution (in total 0.80 mmol, 30 mL) was added to a methanol solution (30 mL) of dpf ligand (0.37 g, 0.80 mmol), and the mixture was refluxed for 3 h. The reaction mixture was concentrated and washed with chloroform. The solvent was evaporated and re-dissolved in 50 °C methanol for recrystallization. The obtained crystals were washed with −20 °C methanol.

Yield: 0.72 g (72% for monomer). IR (KBr): 1657 (st, C=O), 1145 (st, P=O) cm⁻¹. Anal. Calcd for $C_{43}H_{25}Tb_{0.5}Eu_{0.5}F_{18}O_9P_2$: C, 41.48; H, 2.02%. Found: C, 41.40; H, 1.89%.

Preparation of Tb(III)/Eu(III) mixed coordination polymers [Tb,Eu(hfa)₃(dpf)]ₙ (Tb/Eu = 10):

Tb(hfa)₃(H₂O)₂ (0.59 g, 0.73 mmol) and Eu(hfa)₃(H₂O)₂ (59 mg, 0.07 mmol) were dissolved in methanol (30 mL). The Tb(III)/Eu(III)-mixed solution (in total 0.80 mmol, 30 mL) was added to a methanol solution (30 mL) of dpf ligand (0.37 g, 0.80 mmol), and the mixture was refluxed for 3 h. The reaction mixture was concentrated and washed with chloroform. The solvent was evaporated and re-dissolved in 50 °C methanol for recrystallization. The obtained crystals were washed with −20 °C methanol.

Yield: 0.67 g (67% for monomer). IR (KBr): 1657 (st, C=O), 1145 (st, P=O) cm⁻¹. Anal. Calcd for $C_{43}H_{25}Tb_{0.91}Eu_{0.09}F_{18}O_9P_2$: C, 41.39; H, 2.02%. Found: C, 41.36; H, 1.89%.

6.2.4 Crystallography

Single crystals of [Ln(hfa)₃(dpf)]ₙ (Ln=Tb, Gd, Eu) were prepared by slow evaporation of [Ln(hfa)₃(dpf)]ₙ/methanol solution. A colorless block-shaped single crystal of [Ln(hfa)₃(dpf)]ₙ was mounted on a MiTiGen micromesh using Paratone-N. The measurement was performed on a Rigaku R-AXIS RAPID diffractometer using graphite monochromated Mo-K$_\alpha$ radiation. Non-hydrogen atoms were refined anisotropically. Hydrogen atoms were refined using the riding model. All calculations were performed using the Crystal Structure crystallographic software package. CIF data was confirmed by using the checkCIF/PLATON service.

6.2.5 Optical Measurements

UV-Vis absorption spectra were recorded on a JASCO V-670 spectrometer. Emission and excitation spectra were recorded on a HORIBA Fluorolog-3 spectrofluorometer and corrected for the response of the detector system. Emission lifetimes (τ_{obs}) were measured using the third harmonics (355 nm) of a Q-switched Nd:YAG laser (Spectra Physiscs, INDI-50, fwhm = 5 ns, λ = 1064 nm) and a photomultiplier (Hamamatsu photonics, R5108, response time $\leq$1.1 ns). The Nd:YAG laser response was monitored with a digital oscilloscope (Sony Tektronix, TDS3052, 500 MHz) synchronized to the single-pulse excitation. Emission lifetimes were determined from the slope of logarithmic plots of the decay profiles. Emission lifetimes and emission spectra from the range between 100 and 350 K were measured with a cryostat (Thermal Block Company, SA-SB245T) and a temperature controller (Oxford, Instruments, ITC 502S). The emission quantum yields excited at 380 nm (Φ_{tot}) were estimated using JASCO F-6300-H spectrometer attached with JASCO ILF-533 integrating sphere unit ($\varphi = 100$ mm). The wavelength dependence of the detector response and the beam intensity of Xe light source for each spectrum were calibrated using a standard light source. The TL spectra were recorded on a Photonic Hamamatsu PMA-12 Multichannel Analyzer.

6.3 Results and Discussion

6.3.1 Coordination Structures

The crystal structure of [Eu(hfa)$_3$(dpf)]$_n$ is shown in Fig. 6.2a. The Eu(III) ions are coordinated to eight oxygen atoms from three hfa ligands and two bridging ligands. The space group is categorized as non-centrosymmetric C-centered. In order to investigate the geometrical symmetry around Eu(III) ions, the degree of distortion against an 8-coordinated square-antiprismatic structure (8-SAP, point group: D_{4d}) and an 8-coordinated trigonal-dodecahedral structure (8-TDH, point group: D_{2d}) was evaluated on the basis of shape factor S. The observed dihedral angles (δ_i), idealized dihedral angles for square antiprism (θ_{SAP}) and trigonal dodecahedron (θ_{TDH}), calculated measure shape criteria, S_{SAP} and S_{TDH} are summarized (Fig. 6.3; Table 6.1). The S_{SAP} value is much smaller than that of S_{TDH} ($S_{SAP} = 4.98 < S_{TDH} = 12.1$), indicating that the coordination geometry is categorized to be an 8-SAP.

6.3 Results and Discussion   87

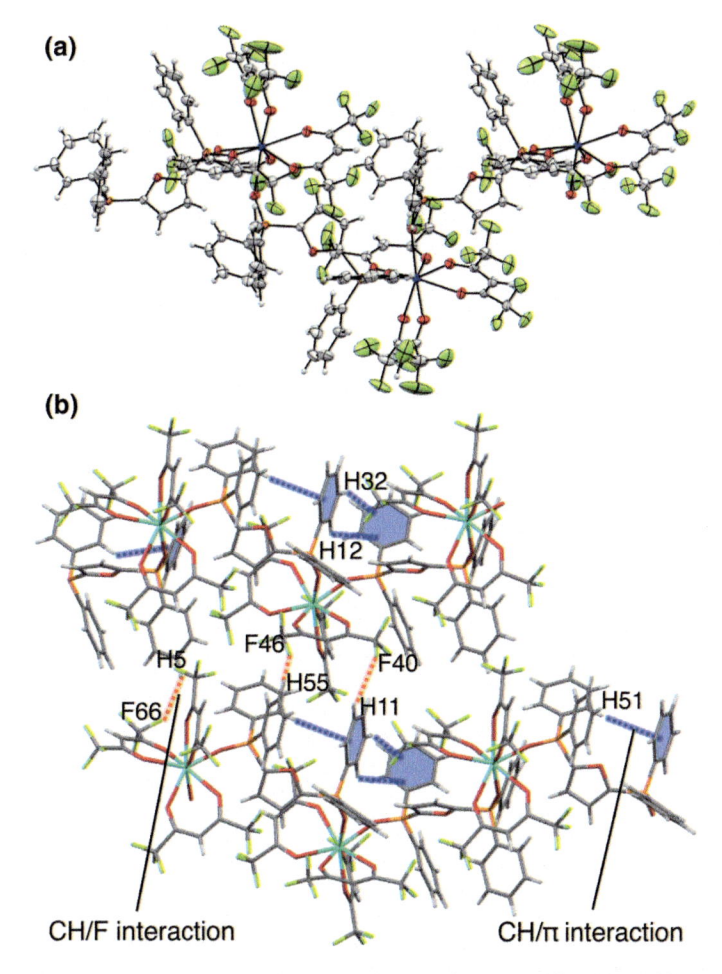

Fig. 6.2 a ORTEP drawing (showing 50% probability displacement ellipsoids) and **b** crystal packing structure focused on intra- and inter-molecular interactions between single polymer chains of [Eu(hfa)$_3$(dpf)]$_n$

Multiple intermolecular CH/F [$d_{CH/F}$ = 2.66 Å (H5/F66), 2.75 Å (H11/F40), 2.93 Å (H55/F46)] and intramolecular CH/π [$d_{CH/\pi}$ = 2.44 Å (H12/π), 2.99 Å (H32/π)] interactions were identified (Fig. 6.2b). On the other hand, intermolecular CH/π interactions, which usually contribute to tight packing structures, were not identified [31]. The relatively weak intermolecular interactions of [Eu(hfa)$_3$(dpf)]$_n$ were attributed to the twisted structure of single polymer chains.

According to the crystal structures observed above and in Chap. 2, Ln(III) coordination polymers linked with polar bridges tend to cancel out the dipole moment by giving a twist to single polymer chains. Coordination polymers with relatively small polar bridges balance the dipoles by intermolecular alternate arrange-

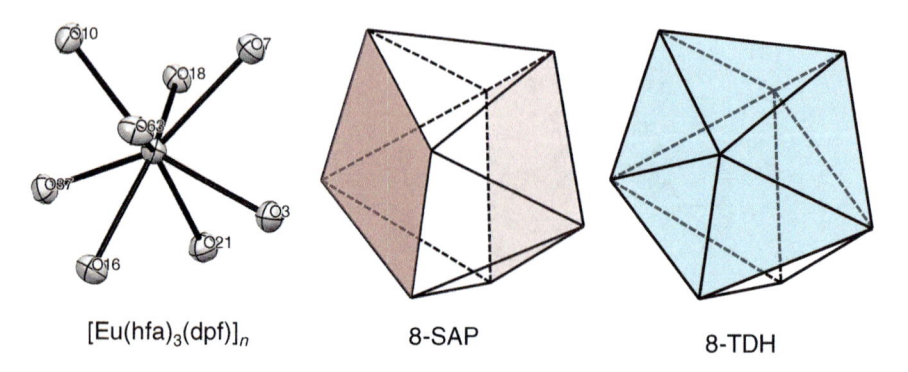

Fig. 6.3 Coordination environments around an Eu(III) ion of [Eu(hfa)$_3$(dpf)]$_n$

Table 6.1 The observed (δ_i), idealized dihedral angles (θ_{SAP}, θ_{TDH}), and calculated S values ($S_{8\text{-SAP}}$, $S_{8\text{-TDH}}$) for [Eu(hfa)$_3$(dpf)]$_n$

	δ_i	SAP			TDH		
		θ_{SAP}	$\delta_i - \theta_{SAP}$	$(\delta_i - \theta_{SAP})^2$	θ_{TDH}	$\delta_i - \theta_{TDH}$	$(\delta_i - \theta_{TDH})^2$
O63–O37	12.23	0	12.23	149.57	29.86	−17.63	310.82
O63–O10	82.10	77.1	5.00	25.00	61.48	20.62	425.18
O63–O16	85.48	77.1	8.38	70.22	74.29	11.19	125.22
O37–O10	77.12	77.1	0.02	0.0004	74.29	2.83	8.01
O37–O16	84.09	77.1	6.99	48.86	61.48	22.61	511.21
O3–O18	3.13	0	3.13	9.80	29.86	−26.73	714.49
O3–O21	81.86	77.1	4.76	22.66	61.48	20.38	415.34
O3–O7	77.71	77.1	0.61	0.37	74.29	3.42	11.70
O18–O7	78.33	77.1	1.23	1.51	76.62	1.71	2.92
O18–O21	73.55	77.1	−3.55	12.60	74.84	−1.29	1.66
O21–O37	56.92	51.6	5.32	28.30	54.51	2.41	5.81
O21–O16	44.33	51.6	−7.27	52.85	48.09	−3.76	14.14
O3–O63	51.09	51.6	−0.51	0.26	47.76	3.33	11.09
O3–O16	49.83	51.6	−1.77	3.13	54.48	−4.65	21.62
O7–O63	53.72	51.6	2.12	4.49	52.57	1.15	1.32
O7–O10	48.83	51.6	−2.77	7.67	52.85	−4.02	16.16
O18–O37	54.33	51.6	2.73	7.45	47.34	6.99	48.86
O18–O10	52.63	51.6	1.03	1.06	52.63	0	0
		$S_{SAP} = 4.98$			$S_{TDH} = 12.12$		

ment. The twisted polymer backbones with disorderly-arranged CF$_3$ substituents in [Eu(hfa)$_3$(dpf)]$_n$ were probably due to the largest ground state dipole moment of dpf among the three bridging ligands (D$_{dpf}$: 5.80 > D$_{dpedot}$: 4.30 > D$_{dpt}$: 1.17 D). The small furyl core also prevented the polymer chains from forming highly ordered packing structures, resulting in the absence of intermolecular CH/π interactions. Therefore, the novel furan-based bridging ligand with a large D value and a small aromatic core well contribute to the formation of a disordered face-to-face arrangement of CF$_3$ substituents in the crystal packing system, leading to a mechanically or a thermodynamically unstable structure.

6.3.2 TL Activity of Eu(III) Coordination Polymers

The ability for generation of TL was ascertained by observing Eu(III)-Characteristic red emission upon grinding of solid Eu(III) coordination polymers (Fig. 6.4a–c). The corresponding space fill models focused on crystal packing structures are also shown in Fig. 6.4d–f. The author observed quite weak TL for [Eu(hfa)$_3$(dpt)]$_n$ under dark, which exhibited a highly ordered arrangement to form a thermodynamically stable structure. The TL activity was clearly high in [Eu(hfa)$_3$(dpf)]$_n$ and [Eu(hfa)$_3$(dpedot)]$_n$, qualitatively, and their red TL can be observed even under daylight. TL intensities of [Eu(hfa)$_3$(dpf)]$_n$ and [Eu(hfa)$_3$(dpedot)]$_n$ were estimated to be approximately 50 times larger than those of [Eu(hfa)$_3$(dpt)]$_n$ (Fig. 6.5). These observations indicate that the ability to generate TL strongly depends on the proportion of face-to-face arrangement of bulky CF$_3$ groups in crystal packing structures. Eliseeva and co-workers also demonstrated strong TL of an Eu(III) coordination polymer [Eu(hfa)$_3$(acetbz)]$_n$ (acetbz: 1,4-diacetoxybenzene) with similar arrangement of CF$_3$ groups between single polymer chains [28]. Therefore, the author also suggests that the mechanical stimuli under grinding generates intermolecular cracking of Ln(III) coordination polymers, resulting in fracture-induced TL.

6.3.3 Photoluminescence (PL) Properties

[Eu(hfa)$_3$(dpf)]$_n$ also exhibited strong red PL under UV irradiation in solid state (Fig. 6.6a, $\lambda_{ex} = 380$ nm). In order to evaluate the PL performance of [Eu(hfa)$_3$(dpf)]$_n$, PL measurements and calculations of photophysical parameters were carried out. The emission bands observed at 578, 591, 613, 650, and 698 nm were assigned to be 4f-4f transitions of $^5D_0 \rightarrow {}^7F_J$ ($J = 0, 1, 2, 3, 4$) in Eu(III) ions. Based on the emission lifetime measurements and calculations, photophysical parameters were compared with those of [Eu(hfa)$_3$(dpt)]$_n$ and [Eu(hfa)$_3$(dpedot)]$_n$ in Chap. 2 (Table 6.2). The emission quantum yield ($\Phi_{tot} = 64\%$) and energy transfer efficiency ($\eta_{sens} = 88\%$) of [Eu(hfa)$_3$(dpf)]$_n$ were higher than those of [Eu(hfa)$_3$(dpt)]$_n$ and [Eu(hfa)$_3$(dpedot)]$_n$. The enhanced radiative rate constant ($k_r = 1.0 \times 10^3$ s^{-1}) in [Eu(hfa)$_3$(dpf)]$_n$ was due to the polar character of bridging ligands and asymmetric coordination geometry without an inversion center around Eu(III) ions. The low-lying ILCT band generated by the stabilized energy level of hfa ligands were responsible for the large Φ_{tot} and η_{sens} values as mentioned for [Eu(hfa)$_3$(dpt)]$_n$ in

90   6 Triboluminescence of Lanthanide Coordination Polymers

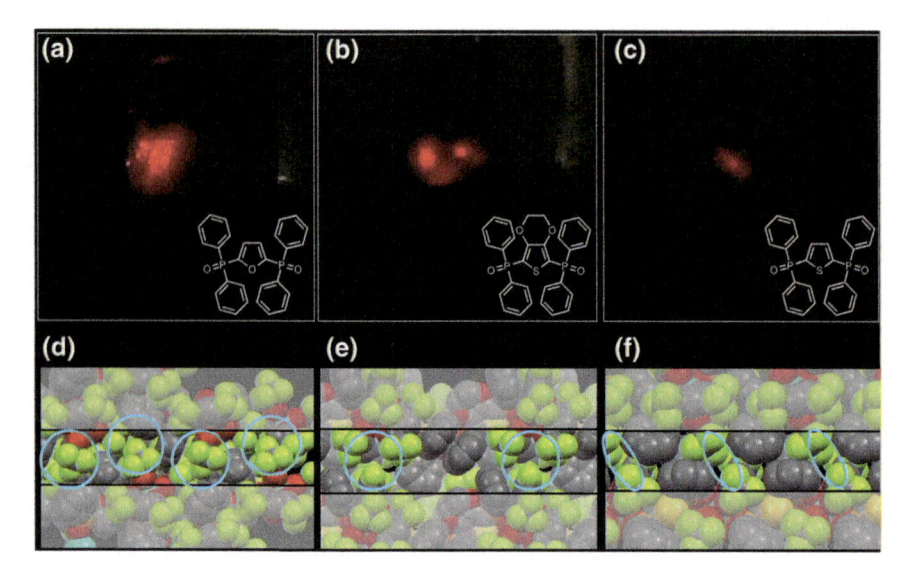

Fig. 6.4 **a–c** Pictures of TL and chemical structures of bridging ligands, and **d–f** corresponding space filling models focused on intermolecular CF_3 arrangement in $[Eu(hfa)_3(dpf)]_n$, $[Eu(hfa)_3(dpedot)]_n$, and $[Eu(hfa)_3(dpt)]_n$

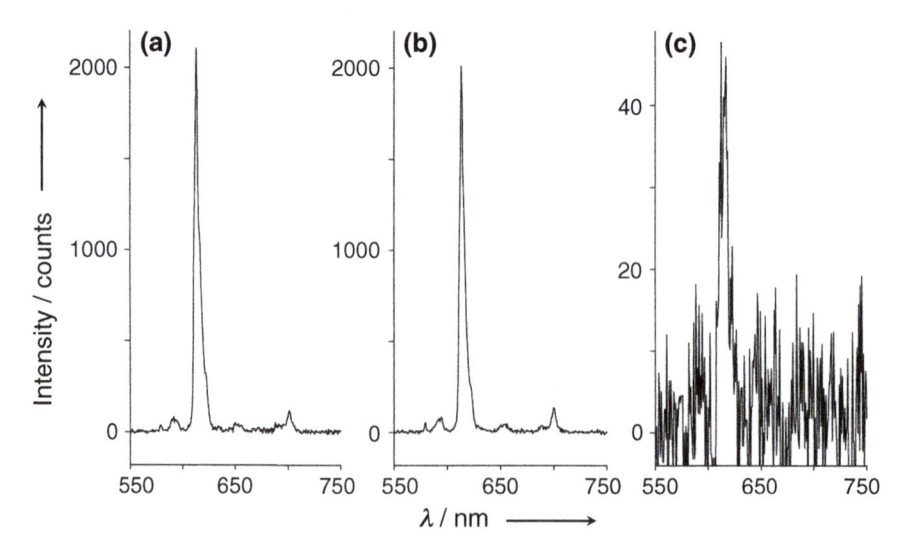

Fig. 6.5 TL spectra of **a** $[Eu(hfa)_3(dpf)]_n$, **b** $[Eu(hfa)_3(dpedot)]_n$, and **c** $[Eu(hfa)_3(dpt)]_n$

Chap. 2. $[Tb(hfa)_3(dpf)]_n$ also exhibited strong green PL with emission bands at 490, 540, 580, 620, and 650 nm, corresponding to 4f–4f transition of $^5D_4 \rightarrow {}^7F_J$ ($J = 6$, 5, 4, 3, 2) in Tb(III) ions (Fig. 6.6b, $\lambda_{ex} = 380$ nm). The intrinsic emission quantum yield (Φ_{ff}) and η_{sens} values were determined to be 88 and 45%, respectively. The

6.3 Results and Discussion

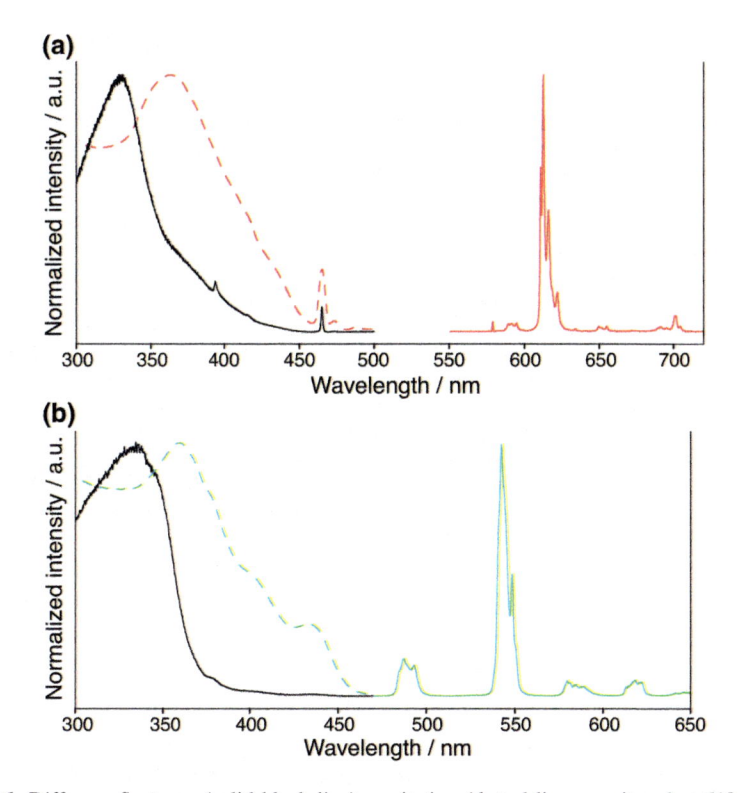

Fig. 6.6 Diffuse reflectance (solid black line), excitation (dotted line, monitored at 613 nm for Eu(III) ions, 545 nm for Tb(III) ions), and emission (λ_{ex} = 380 nm) spectra of **a** [Eu(hfa)$_3$(dpf)]$_n$ and **b** [Tb(hfa)$_3$(dpf)]$_n$ in solid state

Table 6.2 Photophysical parameters of Eu(III) coordination polymers [Eu(hfa)$_3$(X)]$_n$ in solid state

Sample	Φ_{tot}^{a} (%)	Φ_{ff}^{b} (%)	η_{sens}^{c} (%)	τ_{obs}^{c} (ms)	k_r^b (s^{-1})	k_{nr}^b (s^{-1})
[Eu(hfa)$_3$(dpf)]$_n$	64	73	88	0.72	1.0 × 10^3	3.8 × 10^2
[Eu(hfa)$_3$(dpt)]$_n^d$	60	75	80	0.75	1.0 × 10^3	3.3 × 10^2
[Eu(hfa)$_3$(dpedot)]$_n^d$	56	85	66	0.93	9.1 × 10^2	1.6 × 10^2

$^a\lambda_{ex}$ = 380 nm, ^{b}Equations (3–5) in Chap. 2, $^c\lambda_{ex}$ = 355 nm, ^{d}Reference [30]

relatively low η_{sens} value was due to the energy back transfer (BET) from the emissive level of Tb(III) ions to the excited triplet state of hfa ligands [32]. The author considered that the ILCT state which boost energy forward transfer to Eu(III) ions negatively affected Tb(III)-centered emission. Kuzmina and co-workers indicated that appearance of long-wavelength band resulted in an improved energy transfer efficiency for an Eu(III) compound, although for a Tb(III) compound it increased the probability of energy back transfer [33].

6.3.4 PL and TL Properties in Tb(III)/Eu(III) Mixed Systems

Under the condition of grinding, [Eu(hfa)$_3$(dpf)]$_n$ and [Tb(hfa)$_3$(dpf)]$_n$ exhibited strong red and green TL (see Fig. 6.7 for TL and PL spectra of [Tb(hfa)$_3$(dpf)]$_n$). When the same excitation process in TL and PL is assumed, the emission colors of Tb(III)/Eu(III) mixed coordination polymers should be the same under grinding and under UV irradiation. Thus, Tb(III)/Eu(III) mixed coordination polymers were prepared for understanding the relationships between TL and PL properties. The author here considered that the dpf ligand is suitable for Tb(III)/Eu(III) mixed systems, since both [Tb(hfa)$_3$(dpf)]$_n$ and [Eu(hfa)$_3$(dpf)]$_n$ exhibited strong TL and PL with isomorphic crystal structures (Table 6.3).

[Tb,Eu(hfa)$_3$(dpf)]$_n$ (Tb/Eu = 1) was selected to compare the TL and PL efficiencies of Tb(III) and Eu(III) ions under existence of same amount of emission centers. PL colors from Tb(III) and Eu(III) mixed hfa compounds are generally dominated by Eu(III)-centered emission [34–36]. For this reason, [Tb,Eu(hfa)$_3$(dpf)]$_n$ (Tb/Eu = 10) was also selected to tune Tb(III)- and Eu(III)-PL intensity to the same extent. The coordination polymers [Tb,Eu(hfa)$_3$(dpf)]$_n$ (Tb/Eu = 1, 10) were successfully prepared, and the obtained compounds exhibited strong TL and PL under UV irradiation and grinding. The TL and PL spectra were independently recorded using CCD system (Fig. 6.8). The emission intensity of Eu(III) ions is larger than that of Tb(III)

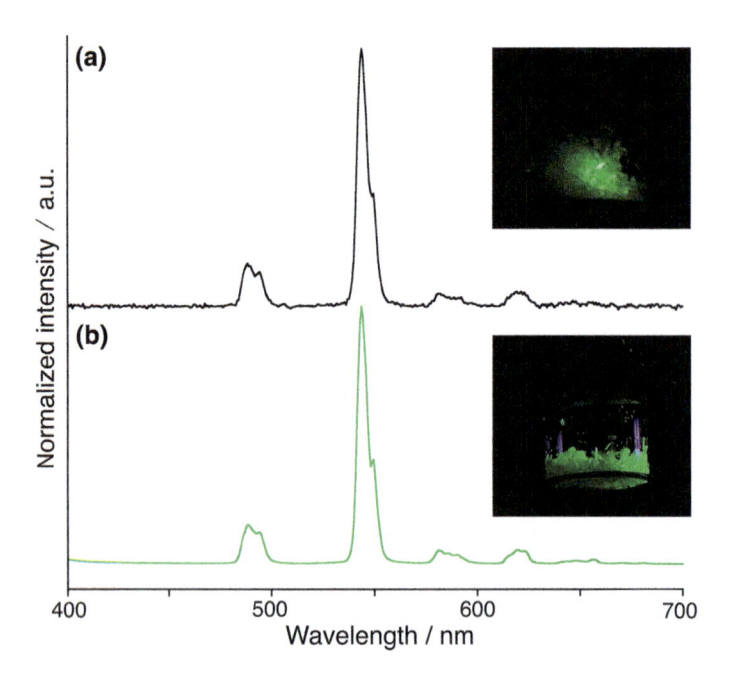

Fig. 6.7 **a** TL and **b** PL spectra of [Tb(hfa)$_3$(dpf)]$_n$ (λ_{ex} = 365 nm, inset shows pictures under grinding and UV irradiation)

6.3 Results and Discussion

Table 6.3 Crystallographic data of the [Ln(hfa)$_3$(dpf)]$_n$ (Ln=Tb, Gd, Eu)

	[Tb(hfa)$_3$(dpf)]$_n$	[Gd(hfa)$_3$(dpf)]$_n$	[Eu(hfa)$_3$(dpf)]$_n$
Chemical formula	C$_{43}$H$_{25}$F$_{18}$O$_9$P$_2$Tb	C$_{43}$H$_{25}$F$_{18}$GdO$_9$P$_2$	C$_{43}$H$_{25}$EuF$_{18}$O$_9$P$_2$
Formula weight	1248.51	1246.83	1241.54
Crystal system	Monoclinic	Monoclinic	Monoclinic
Space group	Cc(#9)	Cc(#9)	Cc(#9)
a (Å)	19.9974(8)	19.9832(7)	19.9767(6)
b (Å)	14.8560(4)	14.8898(5)	14.9113(4)
c (Å)	16.0129(6)	16.0514(5)	16.0652(5)
α (°)	90.000	90.000	90.000
β (°)	97.0396(14)	96.9250(10)	96.9906(9)
γ (°)	90.000	90.000	90.000
Volume (Å)	4721.3(3)	4741.2(3)	4749.9(2)
Z	4	4	4
d_{calc} (g cm^{-3})	1.756	1.747	1.736
Temperature (°C)	−150	−150	−150
μ (Mo K$_\alpha$) (cm^{-1})	16.868	15.938	15.085
max 2θ (°)	55.0	54.8	54.9
Reflections collected	21,629	21,833	22,274
Independent reflections	9599	10,146	9993
R_1	0.0265	0.0266	0.0306
wR_2	0.0605	0.0540	0.0660

$^a R_1 = \sum ||F_o| - |F_c|| / \sum F_o$, $^b wR_2 = [\sum w (F_o^2 - F_c^2)^2 / \sum w (F_o^2)^2]^{1/2}$

ions in the Tb/Eu = 1 compound under UV irradiation, resulting in reddish-orange PL. The PL colors of Tb(III)/Eu(III) mixed coordination polymers are dominated by Eu(III)-centered emission, since the Tb(III)-centered emission is affected by both the BET process and excitation energy transfer from Tb(III) to Eu(III) ions at room temperature [34–36].

Interestingly, the observed TL colors were clearly different from those of PL. The Tb/Eu = 1 compound exhibited yellow TL and reddish-orange PL (Fig. 6.9a), while the Tb/Eu = 10 compound exhibited green TL and greenish-yellow PL (Fig. 6.9b). These spectral differences between TL and PL were the most remarkable among previously reported organic, inorganic, and coordination compounds with TL and PL properties. The TL colors of these coordination polymers might not be explained by the simple excitation mechanisms in PL. Since the TL colors correspond to the Tb(III)/Eu(III) mixture ratios, TL would be dominated by direct excitation of Ln(III) ions; though, the possibility of antenna ligand (hfa)-mediated emission cannot be ruled out.

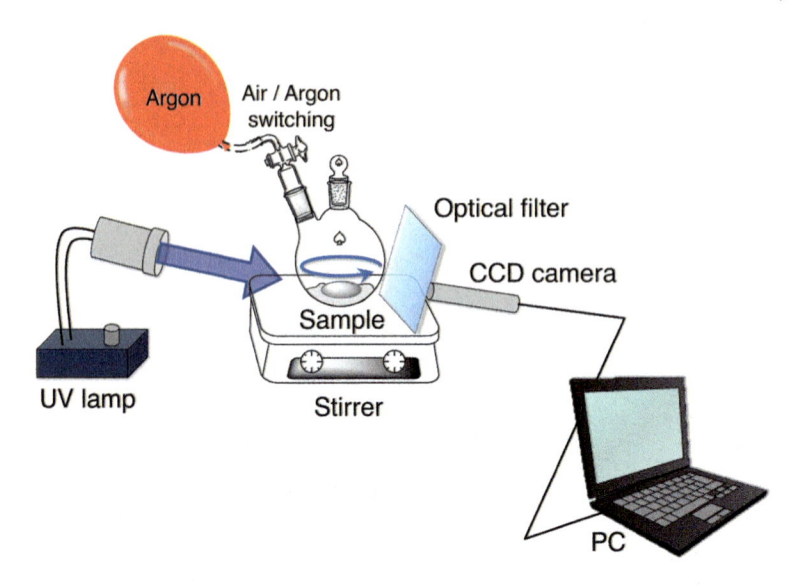

Fig. 6.8 Experimental setups for TL and PL measurements (UV lamp: $\lambda_{ex} = 365$ nm, Stirrer: $v_{rot} = 1500$ rpm, Optical filter: 390 nm longpass, CCD camera: exposure time = 1000 ms, averaging = 2 times)

6.3.5 PL and TL Properties of Gd(III) Coordination Polymers

With the view of the possibility of triplet-excited-state mediated TL, the TL/PL measurements were performed for [Gd(hfa)$_3$(dpf)]$_n$. It is generally known that the Gd(III) $^6P_{7/2}$ level (at 32 100 cm^{-1}) is too high to accept the energy from the triplet excited states of hfa ligands (at 22 000 cm^{-1}). The characteristic blue PL and TL of [Gd(hfa)$_3$(dpf)]$_n$ were observed with identical spectral shapes (Fig. 6.10), indicating the formation of excited states of hfa ligands under grinding. In the case of [Ln(hfa)$_3$(dpf)]$_n$ (Ln=Tb, Eu), hfa-centered emission is quenched through efficient ligand-to-Ln(III) energy transfer, resulting in intense Ln(III)-centered emission.

6.3.6 Atmospheric Dependence of TL and PL

In order to lastly confirm the possibility of TL generated by excited nitrogen gas in an air, the author also measured the TL and PL spectra of [Tb,Eu(hfa)$_3$(dpf)]$_n$ (Tb/Eu = 1) under an argon atmosphere and an air. The spectral difference depending on atmosphere was not observed for both TL and PL (Fig. 6.11). The characteristic blue emission of excited nitrogen gas in the air at around 400 nm was not observed under grinding , too. In addition, the compound exhibited TL even under

6.3 Results and Discussion 95

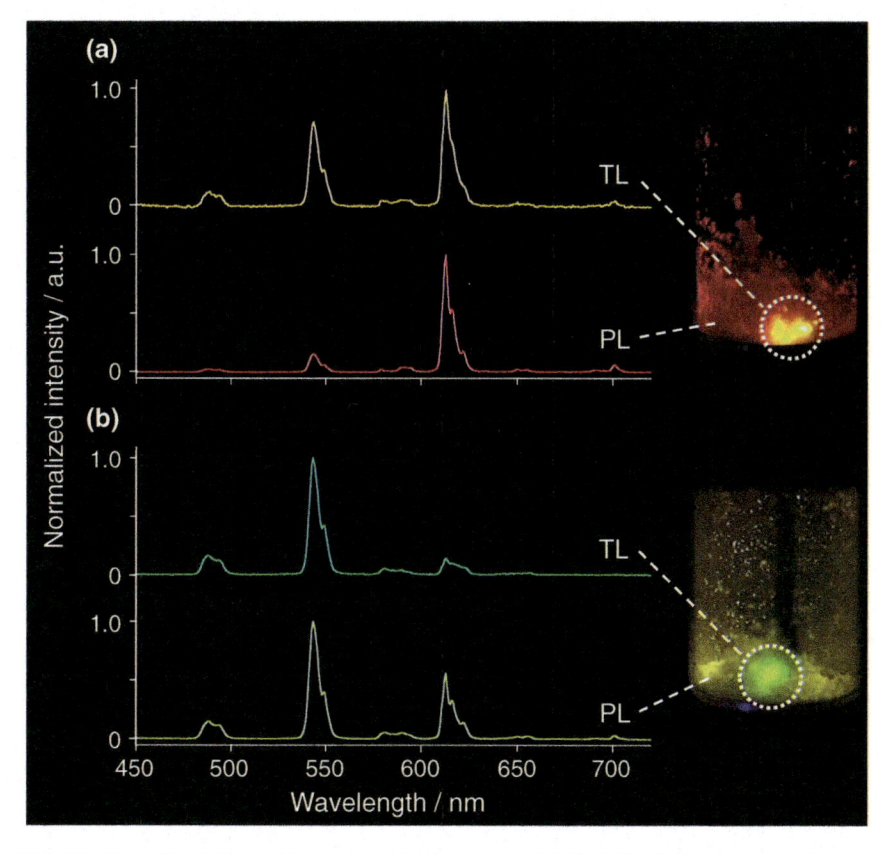

Fig. 6.9 Normalized TL and PL spectra and pictures of **a** [Tb,Eu(hfa)$_3$(dpf)]$_n$ (Tb/Eu = 1) and **b** [Tb,Eu(hfa)$_3$(dpf)]$_n$ (Tb/Eu = 10)

vacuum. Thus, the TL of [Ln(hfa)$_3$(dpf)]$_n$ occur through neither (i) ligand excitation by absorbing the blue emission of excited nitrogen gas nor (ii) energy transfer from excited nitrogen gas to Ln(III) excited state.

6.3.7 Summary of Observed Properties and Further Assumptions

The excitation process of TL in Ln(III) complexes has been basically discussed by ligand-excitation or direct Ln(III)-excitation. Sweeting and Rheingold reported that the charge separation upon cleavage excited the antenna ligands, followed by the formation of Ln(III) excited states via intramolecular energy transfer (ligand-excitation) [27]. Bourhill and co-workers also described Ln(III) excitation by electron bombardment (direct Ln(III)-excitation) [37].

Fig. 6.10 a TL and b PL spectra of [Gd(hfa)$_3$(dpf)]$_n$ (λ_{ex} = 365 nm)

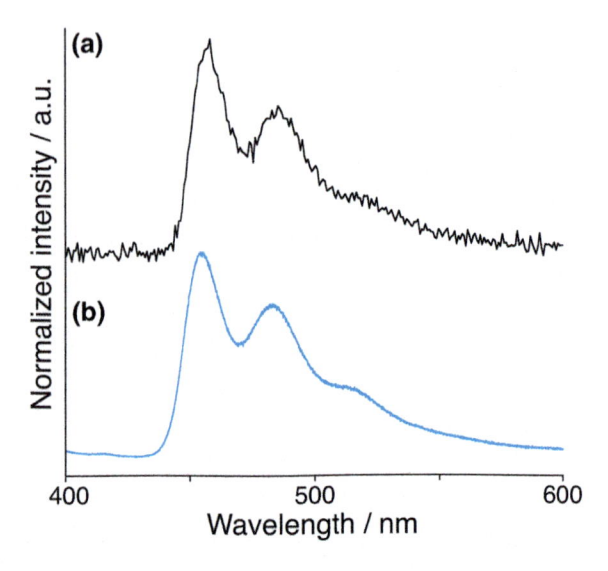

Fig. 6.11 a TL and b PL spectra spectra of [Tb,Eu(hfa)$_3$(dpf)]$_n$ (Tb/Eu = 1) under an argon atmosphere (solid line) and an air (dashed line)

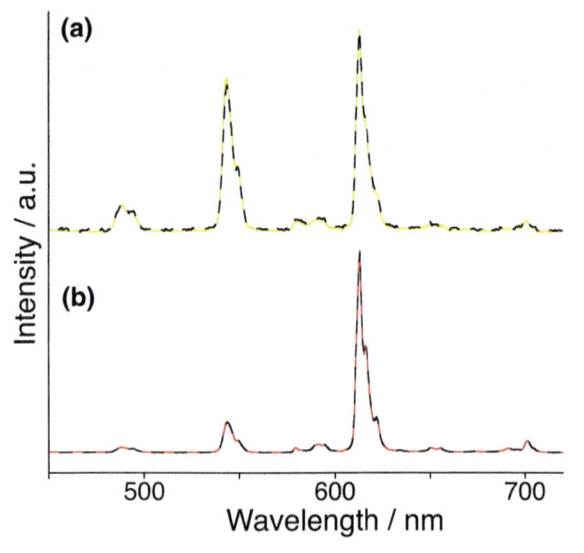

In this study, we observed a large spectral difference between TL and PL. The contribution of Tb(III)-centered emission is larger than that of Eu(III)-centered emission in TL, in contrast to PL. According to the results, the author considers the TL of Ln(III) coordination compounds to be responsible for both the ligand-excitation and direct Ln(III)-excitation (Fig. 6.12), unlike the selective excitation with a specific wavelength (e.g., λ_{ex} = 380 nm) in PL. The photophysical parameters in Table 6.4 also indicate that the efficiency of direct Ln(III)-excitation is higher than that of ligand-excitation under the condition of mechanical grinding. Thus, the reddish-orange PL

6.3 Results and Discussion

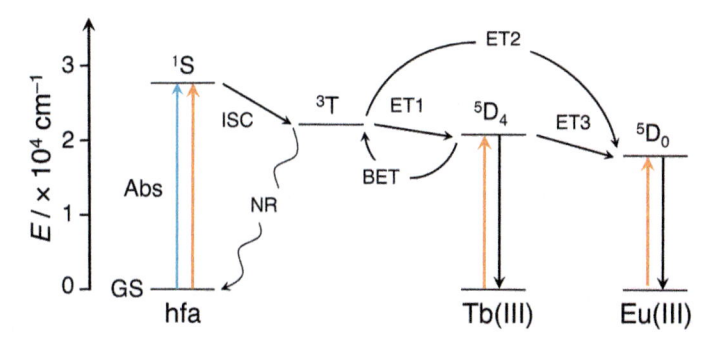

Fig. 6.12 A simplified diagram of excitation processes of TL (orange line) and PL (blue line) in Tb(III)/Eu(III) mixed coordination polymers (Abs: absorption, ET: energy transfer, GS, ground state, ^{1}S: singlet excited state, ^{3}T: triplet excited state, ISC: intersystem crossing, NR: non-radiative relaxation)

Table 6.4 Summary of TL and PL properties of furan-bridged polymers $[Ln(hfa)_3(dpf)]_n$

Sample	Φ_{tot}^a (%)	Φ_{ff}^b (%)	η_{sens}^c (%)	τ_{obs}^c (ms)	PL (color)	TL (color)
$[Eu(hfa)_3(dpf)]_n$	64	73	88	0.72	Red	Red
$[Tb(hfa)_3(dpf)]_n$	40	88^d	45	0.60	Green	Green
$[Tb,Eu(hfa)_3(dpf)]_n$ (Tb/Eu = 1)	–	–	–	0.32 (τ_{Tb})	Reddish-orange	Yellow
$[Tb,Eu(hfa)_3(dpf)]_n$ (Tb/Eu = 10)	–	–	–	0.36 (τ_{Tb})	Greenish-yellow	Green

$^a\lambda_{ex}$ = 380 nm, ^{b}Equations (3–5) in Chap. 2, $^c\lambda_{ex}$ = 355 nm, $^d\Phi_{ff}$ for Tb(III) complex was estimated by $\Phi_{ff,Tb}=(\tau_{obs,RT}/\tau_{obs,100K}) \times 100$, assuming that the τ_{obs} at 100 K is a lifetime without non-radiative process

and yellow TL of $[Tb,Eu(hfa)_3(dpf)]_n$ (Tb/Eu = 1) can be explained by the higher Φ_{ff} of Tb(III) ions ($\Phi_{ff,Tb}$ = 88%) than that of Eu(III) ions ($\Phi_{ff,Eu}$ = 73%). The greenish-yellow PL and green TL in the Tb/Eu = 10 polymer can also be explained by this difference. When considered in the same manner, the Tb(III)/Eu(III) mixed coordination polymers $[Tb,Eu(hfa)_3(X)]_n$ (Tb/Eu = 1, $\Phi_{ff,Tb} < \Phi_{ff,Eu}$) with yellow PL would exhibit orange or red TL.

6.4 Conclusions

Highly TL-active and strong PL Ln(III) coordination polymers were successfully prepared by introducing a furan-based polar bridging ligand. The relationships between crystal structures and TL activity were systematically studied using Eu(III) coordination polymers $[Eu(hfa)_3(dpf)]_n$, $[Eu(hfa)_3(dpedot)]_n$, and $[Eu(hfa)_3(dpt)]_n$ with face-to-face/alternate intermolecular packing structures. The face-to-face arrangement of bulky substituents of the ligands plays an important role in providing strong TL activity. The potential differences between TL and PL were clearly demonstrated by the TL and PL spectra of Tb(III)/Eu(III) mixed coordination polymers for the first time. Based on the observations, the existence of discrete excitation mechanisms for TL and PL phenomena was suggested.

The strategy for constructing strong TL-active compounds would be a key to reveal the ambiguous relationships between luminescence and mechanical stress. These compounds with Tb(III) and Eu(III) ions in particular are also expected to be useful as optical sensors in the field of fluid dynamics and aeronautical engineering, being sensitive to impact, pressure, and temperature to visualize damage and fluid flow on material's surfaces. In addition, compounds with varied emission colors depending on mechanical stress and photo-irradiation are highly attractive as advanced security materials for future identification cards. The results of this work should lead to an understanding of triboluminescence, which has long been a puzzling phenomenon in the field of natural science. More precise and quantitative experiments are expected in the near future to get more insight into the mechanisms of triboluminescence.

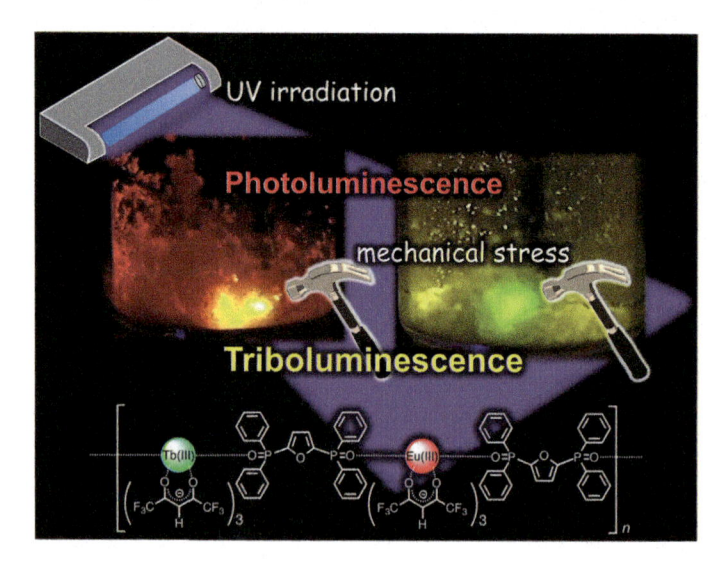

References

1. D.O. Olawale, T. Dickens, W.G. Sullivan, O.I. Okoli, J.O. Sobanjo, B. Wang, J. Lumin. **131**, 1407–1418 (2011)
2. X.D. Wang, H.L. Zhang, R.M. Yu, L. Dong, D.F. Peng, A.H. Zhang, Y. Zhang, H. Liu, C.F. Pan, Z.L. Wang, Adv. Mater. **27**, 2324–2331 (2015)
3. D.O. Olawale, K. Kliewer, A. Okoye, T.J. Dickens, M.J. Uddin, O.I. Okoli, J. Lumin. **147**, 235–241 (2014)
4. R.S. Fontenot, W.A. Hollerman, K.N. Bhat, M.D. Aggarwal, B.G. Penn, Poly. J. **46**, 111–116 (2014)
5. G.P. Williams, T.J. Turner, Solid State Commun. **29**, 201–203 (1979)
6. J.I. Zink, Inorg. Chem. **14**, 555–558 (1975)
7. J.I. Zink, G.E. Hardy, J.E. Sutton, J. Phys. Chem. **80**, 248–249 (1976)
8. G.E. Hardy, J.C. Baldwin, J.I. Zink, W.C. Kaska, P.H. Liu, L. Dubois, J. Am. Chem. Soc. **99**, 3552–3558 (1977)
9. K.F. Wang, L.R. Ma, X.F. Xu, S.Z. Wen, J.B. Luo, Sci. Rep. **6**, 1–9 (2016)
10. A.J. Walton, Adv. Phys. **26**, 887–948 (1977)
11. J.C. Zhang, L.Z. Zhao, Y.Z. Long, H.D. Zhang, B. Sun, W.P. Han, X. Yan, X.S. Wang, Chem. Mater. **27**, 7481–7489 (2015)
12. H. Nakayama, J. Nishida, N. Takada, H. Sato, Y. Yamashita, Chem. Mater. **24**, 671–676 (2012)
13. J. Nishida, H. Ohura, Y. Kita, H. Hasegawa, T. Kawase, N. Takada, H. Sato, Y. Sei, Y. Yamashita, J. Org. Chem. **81**, 433–441 (2016)
14. F. Marchetti, C. Di Nicola, R. Pettinari, I. Timokhin, C. Pettinari, J. Chem. Educ. **89**, 652–655 (2012)
15. C.-W. Hsu, K.T. Ly, W.-K. Lee, C.-C. Wu, L.-C. Wu, J.-J. Lee, T.-C. Lin, S.-H. Liu, P.-T. Chou, G.-H. Lee, Y. Chi, A.C.S. Appl, Mater. Inter. **8**, 33888–33898 (2016)
16. J. Chen, Q. Zhang, F.-K. Zheng, Z.-F. Liu, S.-H. Wang, A.Q. Wu, G.-C. Guo, Dalton Trans. **44**, 3289–3294 (2015)
17. X.-F. Chen, X.-H. Zhu, Y.-H. Xu, S. Shanmuga, Sundara Raj, S. Ozturk, H.-K. Fun, J. Ma, X.-Z. You. J. Mater. Chem. **9**, 2919–2922 (1999)
18. J.P. Duignan, I.D.H. Oswald, I.C. Sage, L.M. Sweeting, K. Tanaka, T. Ishihara, K. Hirao, G. Bourhill, J. Lumin. **97**, 115–126 (2002)
19. K.A. Gschneidner Jr., J.-C.G. Bünzli, V.K. Pecharsky, *Handbook on the Physics and Chemistry of Rare Earths* (Elsevier, Amsterdam, 2005)
20. D.O. Olawale, O.O.I. Okoli, R.S. Fontenot, W.W. Hollerman, *Triboluminescence: Theory, Synthesis, and Application* (Switzerland, Springer Nature, 2016)
21. S.V. Eliseeva, J.-C.G. Bünzli, Chem. Soc. Rev. **39**, 189–227 (2010)
22. J.-C.G. Bünzli, C. Piguet, Chem. Soc. Rev. **34**, 1048–1077 (2005)
23. K. Binnemans, Chem. Rev. **109**, 4283–4374 (2009)
24. S.J. Butler, D. Parker, Chem. Soc. Rev. **42**, 1652–1666 (2013)
25. S. Petoud, G. Muller, E.G. Moore, J. Xu, J. Sokolnicki, J.P. Riehl, U.N. Le, S.M. Cohen, K.N. Raymond, J. Am. Chem. Soc. **129**, 77–83 (2007)
26. S. Petoud, S.M. Cohen, J.-C.G. Bünzli, K.N. Raymond, J. Am. Chem. Soc. **125**, 13324–13325 (2003)
27. L.M. Sweeting, A.L. Rheingold, J. Am. Chem. Soc. **109**, 2652–2658 (1987)
28. S.V. Eliseeva, D.N. Pleshkov, K.A. Lyssenko, L.S. Lepnev, J.-C.G. Bünzli, N.P. Kuzmina, Inorg. Chem. **49**, 9300–9311 (2010)
29. Y. Hasegawa, R. Hieda, K. Miyata, T. Nakagawa, T. Kawai, Eur. J. Inorg. Chem. 4978–4984 (2011)
30. Y. Hirai, T. Nakanishi, Y. Kitagawa, K. Fushimi, T. Seki, H. Ito, Y. Hasegawa, Angew. Chem. Int. Ed. **55**, 12059–12062 (2016)
31. K. Miyata, T. Ohba, A. Kobayashi, M. Kato, T. Nakanishi, K. Fushimi, Y. Hasegawa, ChemPlusChem **77**, 277–280 (2012)

32. S. Katagiri, Y. Hasegawa, Y. Wada, S. Yanagida, Chem. Lett. **33**, 1438–1439 (2004)
33. S.V. Eliseeva, O.V. Kotova, F. Gumy, S.N. Semenov, V.G. Kessler, L.S. Lepnev, J.-C.G. Bünzli, N.P. Kuzmina, J. Phys. Chem. A **112**, 3614–3626 (2008)
34. K. Miyata, Y. Konno, T. Nakanishi, A. Kobayashi, M. Kato, K. Fushimi, Y. Hasegawa, Angew. Chem. Int. Ed. **52**, 6413–6416 (2013)
35. Y. Hirai, T. Nakanishi, K. Miyata, K. Fushimi, Y. Hasegawa, Mater. Lett. **130**, 91–93 (2014)
36. M. Hatanaka, Y. Hirai, Y. Kitagawa, T. Nakanishi, Y. Hasegawa, K. Morokuma, Chem. Sci. **8**, 423–429 (2017)
37. I. Sage, G. Bourhill, J. Mater. Chem. **11**, 231–245 (2001)

Chapter 7
Summary and Outlook

Abstract In this thesis, the author discussed the design of organic bridging ligands to control the entire moieties of assembled Ln(III) coordination compounds for novel photophysical, thermal, and mechanical properties. These molecular designs provide strong luminescence upon ligand excitation (Chap. 2), tunable emission colors (Chap. 3), glass-formability (Chaps. 4 and 5), and mechanical stress-sensitivity (Chap. 6) depending on the geometrical structures of organic bridging ligands and incorporated Ln(III) ions. In this chapter, a summary of the results, possible applications, and future perspectives are presented.

Keywords Lanthanide · Complex · Coordination polymer · Luminescence

7.1 Summary

The author aimed to systematically control the assembled structures of Ln(III) coordination compounds to achieve novel photophysical, thermal, and mechanical properties. The design of a series of organic bridging ligands to control the entire moieties of Ln(III) complexes and to exploit the advantages of strong luminescent Ln(hfa)$_3$ units is described in this thesis.

In Chap. 1, a brief history of luminescent materials was given. The scientific background of the development of luminescent lanthanide complexes and achievements such as strong luminescence and thermal stability were described.

A novel packing system, "coordination zippers," for improvement of the photosensitization efficiency was introduced in Chap. 2. Dense-packed structures were provided by bent-angled (hetero)aromatic bridging ligands, and characteristic low-lying absorption bands that arose from the stabilized energy level of antenna ligands were observed. Higher energy transfer efficiency ($\eta_{sens} = 80\%$) and higher thermal stability (decomposition point: 322 °C) was achieved using a thiophene-based bridge to form inter-molecular CH/F and CH/π hydrogen bonds.

In Chap. 3, energy transfer between Ln(III) ions was systematically investigated using biphenylene-bridged coordination polymers with various Tb(III)/Eu(III) mixture ratios. The energy transfer efficiencies were found to depend on the mixture ratios, which varied from 99 to 8.2% at room temperature (Tb/Eu = 1–500). Unusual negative energy transfer efficiency was estimated for Tb/Eu = 750, indicating the different concentration quenching processes of Tb(III)- and Eu(III)-centered emission.

In Chap. 4, molecular designs for amorphous and luminescent Ln(III) complexes were reported. Ortho-, meta-, and para-substituted phenylene-bridging ligands with ethynyl groups were prepared. Single crystal X-ray analyses and DFT optimizations revealed the formation of dimer, trimer, and polymer structures. The glass transition properties were found to depend on the moiety of bridging ligands (T_g = 25–65 °C). The ethynyl groups as well as pseudo-C_3 structures contributed to the formation of a stable amorphous solid.

Systematic construction of Ln(III) coordination glasses using bent-angled bridging ligands with ethynyl groups was described in Chap. 5. A temperature-sensitive Tb(III)/Eu(III) mixed coordination glass was prepared following the strategy described in Chaps. 3 and 4. Green, yellow, orange, and red photoluminscence (PL) in a glassy state was successfully observed depending on the temperature.

In Chap. 6, triboluminescence (TL) of Ln(III) coordination polymers was studied. A series of strong TL compounds was prepared by introducing an inter-molecular face-to-face arrangement of CF_3 groups, and relationships between packing structures and TL activity were suggested. In addition, the excitation process of TL from both organic ligands and Ln(III) ions was proposed on the basis of TL/PL measurements of a Gd(III) coordination polymer and Tb(III)/Eu(III) mixed coordination polymers. These features would provided the large spectral difference between TL and PL in Ln(III) mixed systems.

7.2 Outlook

The photophysical, thermal, and mechanical properties of Ln(III) coordination compounds can be controlled by organic bridging ligands. The enhancement of intermolecular interactions leads to thermal stability, and the suppression of intermolecular packing is also advantageous for the formation of glassy materials. In addition, intermolecular disordered arrangement of substituents can trigger stress-sensitive luminescence.

The novel designs of bridging ligands have further potential for the development of functional materials such as low temperature-sensitive luminescent dyes for ice-coating sensing on airframes, thermally stable luminescent Ln(III) coordination glasses for EL devices, and NIR-PL/VIS-TL coordination polymers for hyper-advanced security marking techniques. These "additional" features are attractive for next-generation functional luminescent materials as well as the obvious importance of the strong luminescence.

7.2 Outlook

The author hopes that the molecular designs described in this book break new ground for functional materials with strong luminescence and additional physical properties arising from specific assembled structures. The knowledge about the relationships between photoluminescence and other physical properties is also expected to provide scientific understandings of fluid dynamics in unsteady flows, thermophysics in glass transitions, and photoscience in triboluminescence.

Curriculum Vitae

Yuichi Hirai, Ph.D.
Former affiliation address:
Graduate School of Chemical Sciences and Engineering, Hokkaido University
Kita 13-jo Nishi 8-chome, Kita-ku, Sapporo, Hokkaido, Japan
Web: http://www.eng.hokudai.ac.jp/labo/amc.

Career

- The Japan Society for the Promotion of Science (JSPS) Research Fellow (DC2, 2016–2017).
- The Japan Society for the Promotion of Science (JSPS) Research Fellow (PD, 2017–2018).

Education

- Doctor of Engineering, Hokkaido University, 2014–2017
  (Supervisor: Prof. Yasuchika Hasegawa).
- Master of Engineering, Hokkaido University, 2013–2014
  (Supervisor: Prof. Yasuchika Hasegawa).
- Bachelor of Engineering, Hokkaido University, 2009–2013
  (Supervisor: Prof. Yasuchika Hasegawa).

Award

- 14th Chitose International Forum on Photonics Science & Technology (CIF14), Best Poster Award.
- 25th Meeting on Photochemistry of Coordination Compounds, Poster Award.

- 3rd Chemical Society of Japan (CSJ) Chemistry Festa, Poster Presentation Award
- 15th Chitose International Forum on Photonics Science & Technology (CIF15), Poster Award.
- The 67th Conference of Japan Society of Coordination Chemistry (JSCC), Poster Award.
- The 98th Chemical Society of Japan (CSJ) Annual Meeting, Student Presentation Award.
- International Symposium on Lanthanide Coordination Chemistry (ISLCC2016), Best Poster Award.
- 6th Chemical Society of Japan (CSJ) Chemistry Festa, Poster Presentation Award.
- 28th Rare Earth Research Conference (28th RERC), Best Poster Presentation.

Printed by Printforce, the Netherlands